PREPARING SCIENTIFIC ILLUSTRATIONS

Springer
New York
Berlin
Heidelberg
Barcelona
Budapest
Hong Kong
London
Milan
Paris
Santa Clara
Singapore
Tokyo

Mary Helen Briscoe

PREPARING SCIENTIFIC ILLUSTRATIONS

A Guide to Better Posters, Presentations, and Publications

Second Edition

With 206 illustrations

Springer

Mary Helen Briscoe
Word and Image
250 Ashbury
San Francisco, CA 94117, USA

Library of Congress Cataloging in Publication Data

Briscoe, Mary Helen
Preparing scientific illustrations: a guide to better posters, presentations, and publications/Mary Helen Briscoe.—2nd ed.
p. cm.
Includes bibliographical references and index.
ISBN 0-387-94581-4 (pbk.: alk. paper)
1. Scientific illustration. I. Title.
Q222.B75 1995
502'.2—dc20 95-34366

Printed on acid-free paper.

Production managed by Princeton Editorial Associates and supervised by Natalie Johnson; manufacturing supervised by Joseph Quatela.
Typeset by Princeton Editorial Associates, Princeton, NJ.
Printed and bound by Hamilton Printing Co., Rensselaer, NY.
Printed in the United States of America.

9 8 7 6 5 4 3 2 1

ISBN 0-387-94581-4 Springer-Verlag New York Berlin Heidelberg

In memory of Julius H. Comroe, Jr.,
without whom I would not have started,
and in gratitude to Wayne,
without whom I would not have finished.

ACKNOWLEDGMENTS

For the first edition of this book, I was and still am indebted to the Cardiovascular Research Institute (CVRI) and its director, Dr. Richard J. Havel. Many members of CVRI were particularly helpful with their time and advice, and with the loan of books and figures. I am especially grateful to Dr. Norman Staub and the students and faculty of his classes in The Art of Lecturing—Dr. Leonard Peller, Dr. Richard Bland, Dr. Donald McDonald, Mr. Paul Graf, Ms. Rolinda Wang, Dr. Jen Tsi Yang, and Dr. Wendy Wu—and to all who gave permission to use figures for this book.

Ms. Ilse Sauerwald of the Department of Growth and Development took time to describe photographic processes and to develop and print some of the photographs. Dr. Geri Chen and Mr. David Hardman patiently demonstrated and explained the process of making, photographing, and printing gels.

From the University of California, San Francisco, campus, Dr. Hugo Martinez of the Department of Biochemistry and Biophysics helped to explain sequences and maps. Ms. Jori Mandelman, Dr. Martinez's assistant, generated and explained figures showing sequencing and mapping sections.

Ms. Leslie Taylor of the Department of Pharmacy demonstrated molecular models generated in the computer graphics laboratory. Ms. Julie Newdoll, artist for the Department of Biochemistry and Biophysics, was also helpful in explaining and demonstrating molecular models.

The advice, both scientific and literary, of Dr. Wayne Lanier has been invaluable, as was the astute and painstaking proofreading of Ms. Lilly Mendelson Urbach.

For this second edition, I add my thanks to research biochemist Clive Pullinger, of the Cardiovascular Research Institute, University of California, San Francisco, for his helpful and detailed comments on the first edition and for his demonstration of sequencing software. I am also indebted to Dr. Peter Callen of the Department of Radiology, University of California, San Francisco, for his demonstration of technologies for digitizing pictures.

M.H.B.

Contents

1 INTRODUCTION

The purpose of this book is to help scientists make effective choices by pointing out some principles of good visual communication and by showing examples of good and bad choices.

It is now possible for a cardiologist to transfer a digitally produced echocardiograph into a photo manipulation program where labels may be changed or deleted, a title may be added, selective parts of the echo may be chosen, the image may be enlarged or reduced, and color may be added. Directly from the screen, the image may be made into a slide or a printed page or saved to a disk, which may be sent away for publication.

Thus the possibilities for data handling by computer are marvelous and fascinating. The resulting choices presented seem endless and are often confusing. It becomes hard to remember that the point of these choices is to inform as clearly, quickly, and easily as possible.

Presented with such an array of tools and choices, it is easy to make the wrong choice, which sends an unintended message or obscures information. The following figure is a case in point:

OBSCURE INFORMATION

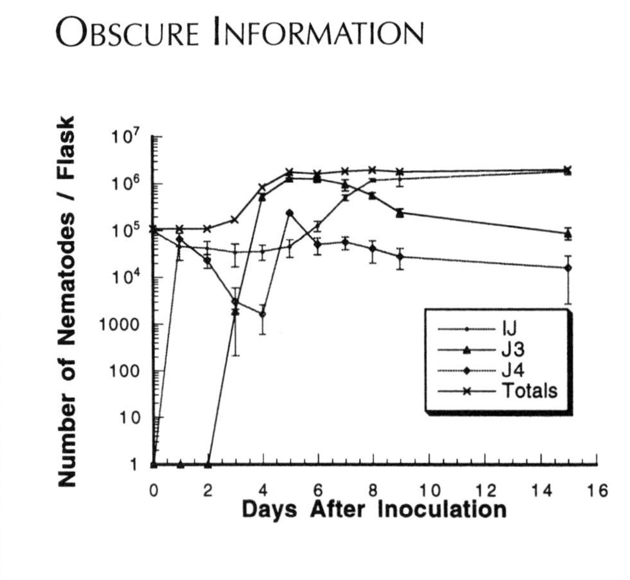

In this figure, four sets of data are plotted from an in vitro culture inoculated with infective juvenile (IJ) nematodes. The figure purports *to show population changes in three stages of development from IJ to 4th- or 3rd-stage juveniles (J3 and J4). Instead it is a visual thicket of tangled information. It is so cluttered with information that the development pathways are meaningless.*

In such a complex figure, it is difficult to know what is important. The eye is drawn first to the shadowed boxed legend. Thus bold large labels and thick axis lines draw attention from the data.

The graph below shows some improvement.

CLEARER INFORMATION

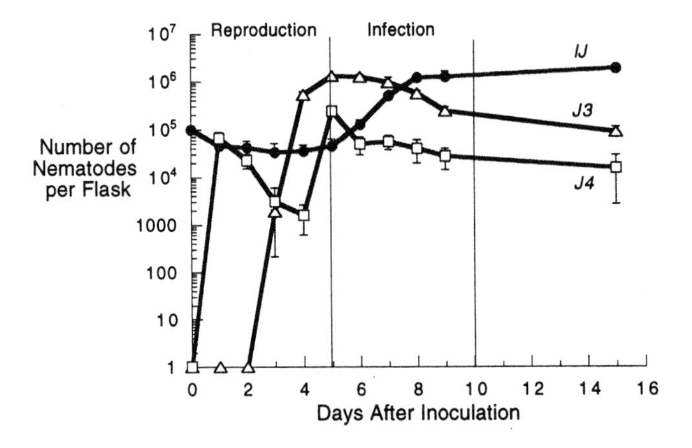

Data for totals are left out because they do not contribute to the message. Symbols for the juveniles are larger, lines connecting them thicker. The axis lines and labels are thinner, label size is in better proportion, the legend is removed, and curves, directly labeled. These small changes in the manner of presentation convey a cleaner, more focused and accurate message.

But still the message may not be clear. Although the preceding graph shows the dynamic changes in the kinds of juveniles over time, it does not show the shifting changes in populations. This is better expressed in the graph below.

STILL CLEARER

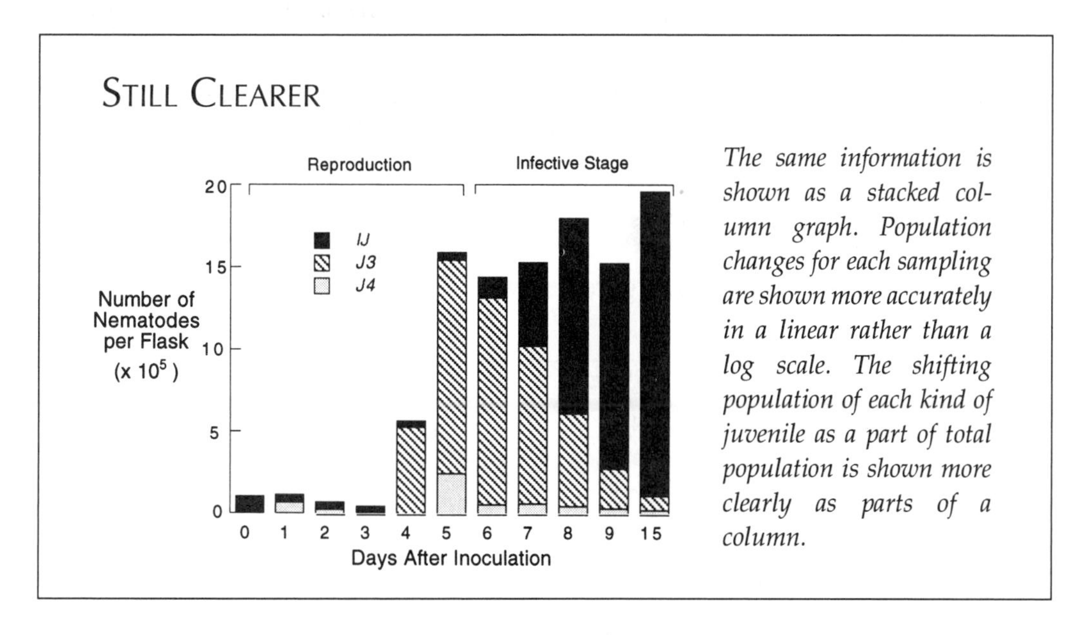

The same information is shown as a stacked column graph. Population changes for each sampling are shown more accurately in a linear rather than a log scale. The shifting population of each kind of juvenile as a part of total population is shown more clearly as parts of a column.

To communicate well visually, the scientist must first know what aspect of findings he or she wants to convey. Then the *manner* of conveying that aspect must be visually accurate and carefully considered and planned.

LEARNING TO COMMUNICATE VISUALLY

Communication, or the sharing of information, is often defined in terms of words. When an investigator writes a paper or prepares to give a talk, he or she first thinks of the words to be used, not pictures. However, because so many of us learn through pictures, visual communication is an important, sometimes essential, and certainly effective way to inform.

Using Words

Below is an example of text describing a process.

> The life cycle of the nematode, *Neoaplectana carpocapsae,* begins with the infective juvenile (IJ) stage. The IJs infect insects, usually the larvae, by being eaten or by penetrating some other body orifice. Once inside the insect gut, the IJs penetrate the gut wall and enter the haemocoel. They begin the first-generation reproductive cycle by rapidly growing and developing into the fourth juvenile stage. The fourth-stage juveniles grow, molt, and develop into adults. Adult males and females mate

> and the first-generation giant females produce as many as 10,000 eggs. The eggs develop into first-stage juveniles, which grow, molt, and become second-stage juveniles. Here the nematodes have a choice of repeating the reproductive cycle by growing and molting into third-stage juveniles, or by entering the infective cycle. This fork in the developmental pathway is controlled by population and nutrition. In an underpopulated host larva, development is toward the reproductive cycle. In a crowded host larva, the second-stage juveniles develop into IJs. The IJs migrate from the exhausted larval host and seek new hosts to infect.

How long did it take to read this? Is it clear? Was reading it an easy, rewarding experience? Could the process be visualized as it was read?

Using Pictures

Compare the text above with the figure below.

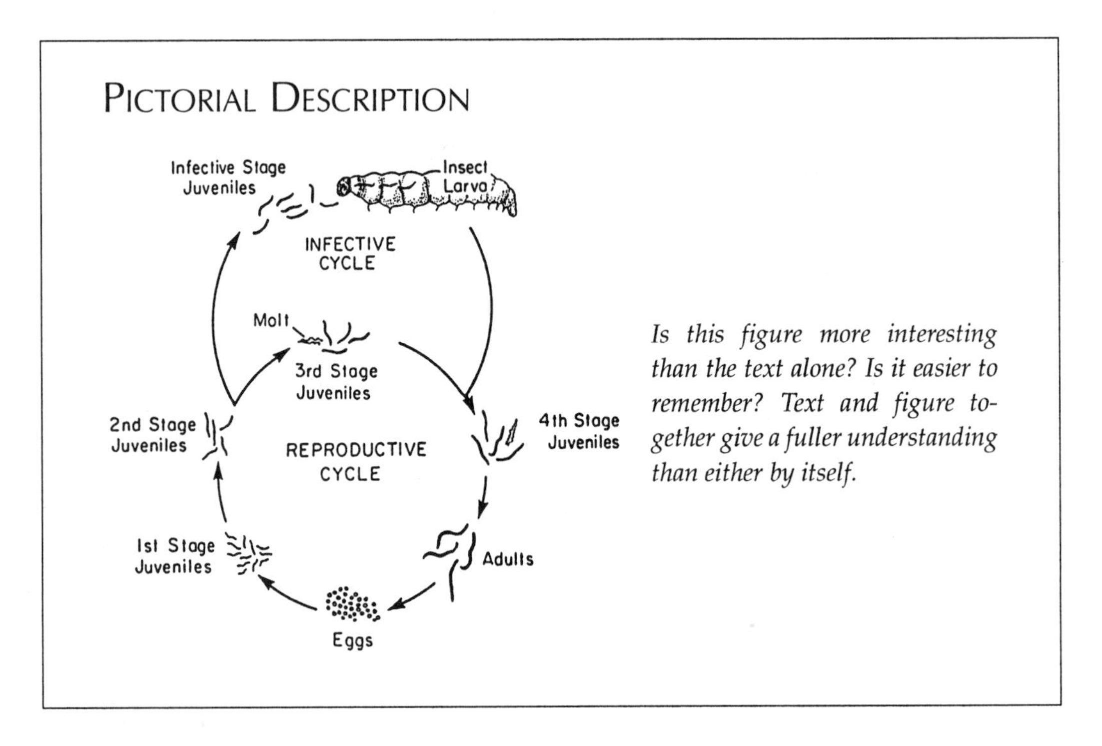

Is this figure more interesting than the text alone? Is it easier to remember? Text and figure together give a fuller understanding than either by itself.

An excellent scientist may be famous for wearing unmatched socks. Another may be acutely observant in the laboratory but could not describe the color of the Golden Gate Bridge. All of us have examples of this phenomenon in and out of science. It is safe to say that we do not see most of the details of our own lives. We are visually unaware.

This book is based on the view that clarification of ideas by visual means, by illustration, is not only effective but can be learned by even the most visually inarticulate.

An Intention to Communicate

In science, accuracy is vital. So also is a genuine intent to inform. Too often a stated intent to inform is belied by the other messages that communicate all too clearly, such as

- Nobody is going to understand.
- See how much I know.
- I am too busy with important work to take pains with this.
- It is up to you to figure this out.

It is essential that the message be as accurate as the calculations. The scientist must make a commitment to inform. Be interested in your own information if you want anyone else to find it interesting. Concentrate on finding ways to make information clear and interesting to others.

Good Illustrations

A good illustration can help the scientist to be heard when speaking, to be read when writing. It can help in the sharing of information with other scientists. It can help to convince granting agencies to fund the research. It can help in the teaching of students. It can help to inform the public of the value of the work.

- A good illustration addresses the viewer appropriately. The kind and number of illustrations, the amount of information conveyed, and the level of simplification or complexity of the illustrations should be designed for a specific viewing audience.
- A good illustration supports, enhances, and emphasizes the purpose and the words.
- A good illustration is suitably designed for the intended medium: paper, slide, poster. It is of a legible size, has concise and consistent labels, and uses clear forms and symbols.
- A good illustration has visual impact and interests and attracts the viewer.

Planning the Figure

Good ideas do not communicate themselves. Ideas must be organized. Highly complex ideas need to be clarified and simplified whereas diffuse data may benefit from being combined. Ideas and data must be made interesting and comprehensible to those not familiar with them.

Thorough planning is essential. Well-planned and meticulously executed figures are a powerful tool for communication. Figures should be given the same consideration as words.

Thoughtful planning can save time and money and result in easily understood and interesting figures. Planning is perhaps the most important part of an illustration. The most careful drawing and the most expensive execution will not compensate for poor planning. We have all had the experience of being stunned by a blazing color, high-tech slide that leaves us baffled when we try to understand its content.

When to Use a Figure

- Use a figure when words will not suffice. Photomicrographs, tracings (as from electrocardiographs), gels, and anatomical drawings are essential for documentation and speed of understanding.
- Use a figure for verisimilitude. Photographs, gels, and tracings are necessary proofs of research findings.
- Use a graph or table to provide the audience with the tools for critical analysis of the work.
- Use a figure to clarify, simplify, and summarize information.
- Use a figure to emphasize information.
- Use a figure to provide background information. Even the sharpest audience appreciates a reminder of what it may already know.

How Much Information

The most common disaster in illustrating is to include too much information in one figure. Too much information in an illustration confuses and discourages the viewer.

- The amount of information is determined by the medium. Posters are limited by size; slide viewing is limited by time; the printed figure is limited by space and money considerations.
- The amount of information to be presented is determined by the intended viewer. Some viewers have the background to understand more complexities in a subject. Others will comprehend more from several simple figures.
- The amount of information is determined by the novelty or controversial content of the information. A series of figures with minimal information that build ideas gradually is most effective for this. (See pages 117–119 in Chapter 8, "Slides.")

Plan for the Viewer

- The viewer's background—such as age, education, and profession—will determine the level of complexity of the figures.
- Some viewers will require more background and explanatory figures. For most viewers, drawings and diagrams are more helpful than words and tables.

Plan for the Medium

- If the figure will be used in a journal paper, instructions to authors should be followed (usually included in one issue per volume). Because these instructions generally have very few guidelines for the presentation of figures, look through the journal to see the style of figures in papers that have been accepted. (See Chapter 7, "The Journal Figure.")
- For lectures, consider the size of the room and the kind of projection equipment available. Check that the equipment is working. It is best to project slides ahead of time to avoid the embarrassment of upside-down and backward slides. (See Chapter 8, "Slides.")
- For poster sessions, know the dimensions of the space available. Know the size of the meeting, the location, and the layout of the poster hall, the breadth of your audience's interests, the quality of the lighting, and the amount of space around and between posters. (See Chapter 9, "Posters.")

Plan Integral Figures

- Keep in mind that each figure is a part of a whole context. It must relate to the other parts of the whole.
- Be consistent in format, size, and terminology.
- Avoid repetition except to emphasize.
- Try to create a smooth, logical flow of information with the figures. Often the essence of the information will be clear from the figures alone, and extensive explanations are not needed.

Plan For Sufficient Time and Money

Vast amounts of time and money are spent in planning and carrying out experiments. Do not sacrifice this effort by failing to apply the same attention to the presentation of the results.

- Allow time to think carefully about the rough figure. Discuss the figure with others.
- Allow time to revise the rough figure, perhaps more than once.
- Allow time for changes and corrections in the final figure.
- Plan to spend more time and money on drawings and complex figures.
- Allow sufficient time and money for the necessary photography, enlargements, and mounting of the final figure.

Plan to Communicate Visually

In planning an illustration, stand away from the information and imagine the viewer. Think of the information in terms of what it will look like to another. Find different ways to present the information that will make it easy for the viewer to understand. Translate the information into visual form that shows ideas accurately and attracts the viewer.

2 Drawings and Diagrams

There are many ways to illustrate scientific and medical information. Sometimes there will be a choice. For instance, should a drawing or a photograph be used to show a laboratory set-up? Should a graph or a table be used to illustrate the data?

In this chapter and the next four chapters are some examples of kinds of illustrations and their appropriate use. "Appropriate" here means fitting the needs to the intended goal. To choose the appropriate illustration, the purpose, audience, and medium must be understood.

When the French artist Edgar Degas stated, "In a single brush stroke, we can say more than a writer in a whole volume," he might have added, "We can say it compellingly and interestingly, too." An appropriate drawing can eliminate pages of text; it can give insights; it can be a vivid and memorable part of the information. In addition to describing succinctly, it can add vigor to the presentation.

On the following pages are examples of a few of the ways in which drawings may be used.

ANATOMICAL DRAWING

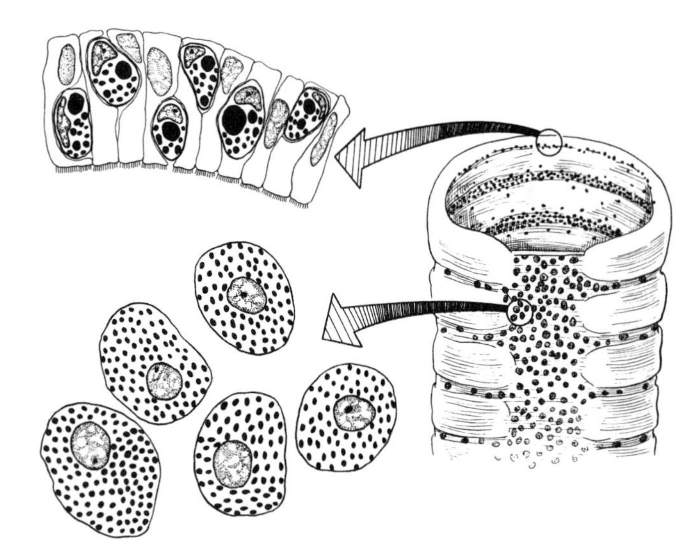

A drawing can simplify anatomical complexities. This line drawing juxtaposes tracheal anatomy with an exploded view of the tracheal cells. This is one advantage a drawing has over a photograph.

MICROANATOMICAL DRAWING

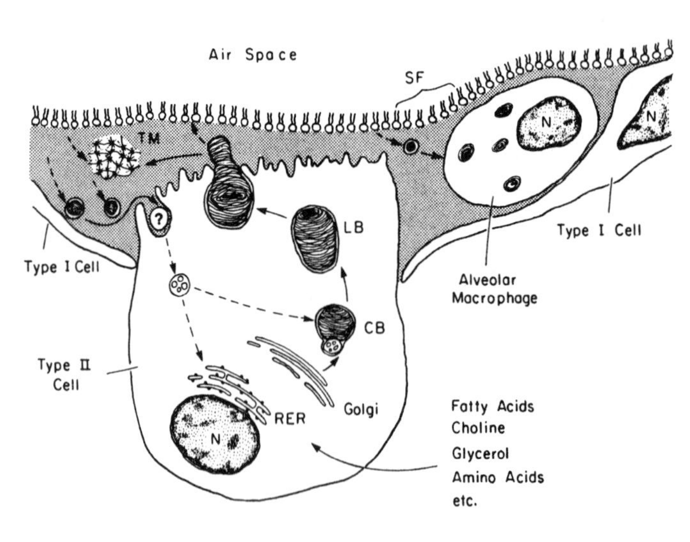

This drawing summarizes elements found in several photomicrographs. By simplifying, it emphasizes essential points of microanatomy and clearly identifies those elements. The arrows indicate dynamic movement of the pathways.

PATHOLOGY DRAWING

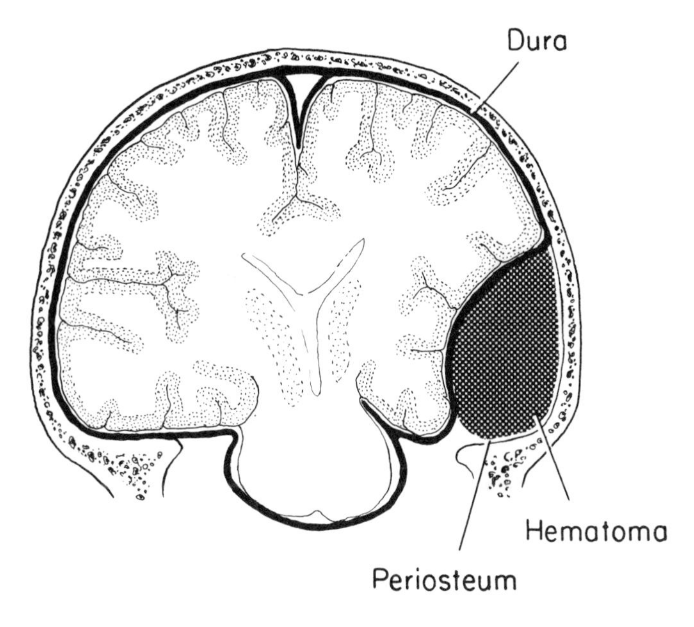

A drawing is an effective way to emphasize infomation. Here, attention is called to a cerebral hematoma. The skull and cerebrum are simply outlined, but the areas of interest are darker and textured.

SURGICAL DRAWING

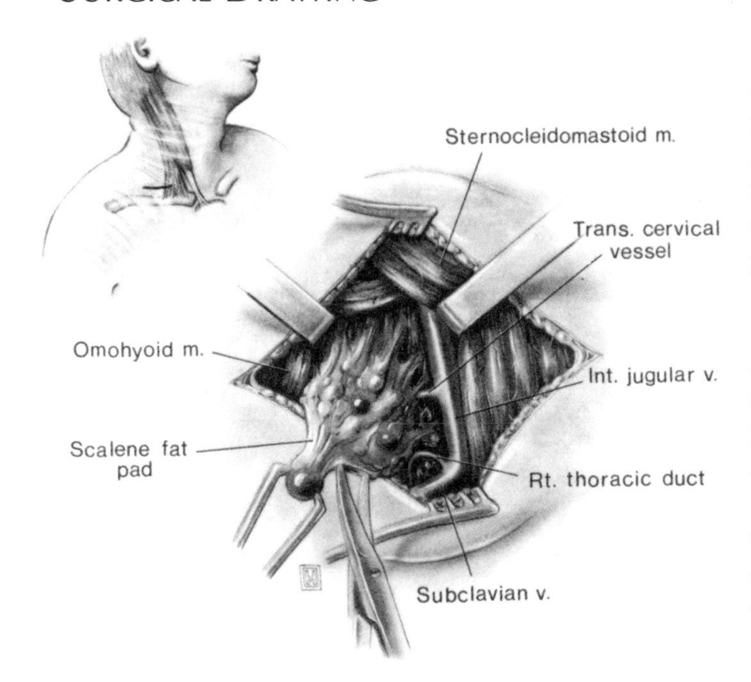

A drawing of surgery removes inessentials. It eliminates distracting background and simplifies and emphasizes information. This drawing shows the isolation of the scalene node more clearly and simply than a photograph. In the head and neck view, the muscles of the neck are ghosted under the skin and the line of incision is shown.

PROCEDURAL DRAWING

This drawing isolates the essential elements of a process. Unrealistic proportions are useful to emphasize one element over another.

DIAGRAM OF BIOCHEMICAL PATHWAYS

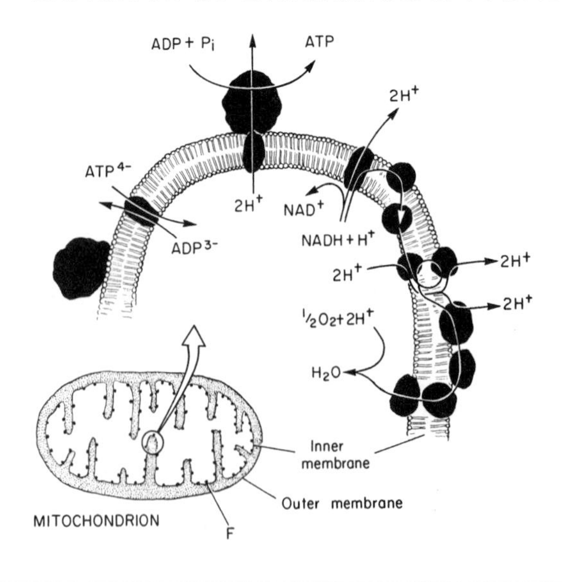

This exploded view of mitochondrial membranes shows dynamic movement of elements across the membranes. Drawings or diagrams are an excellent way to present hypotheses, theories, and imaginative ideas.

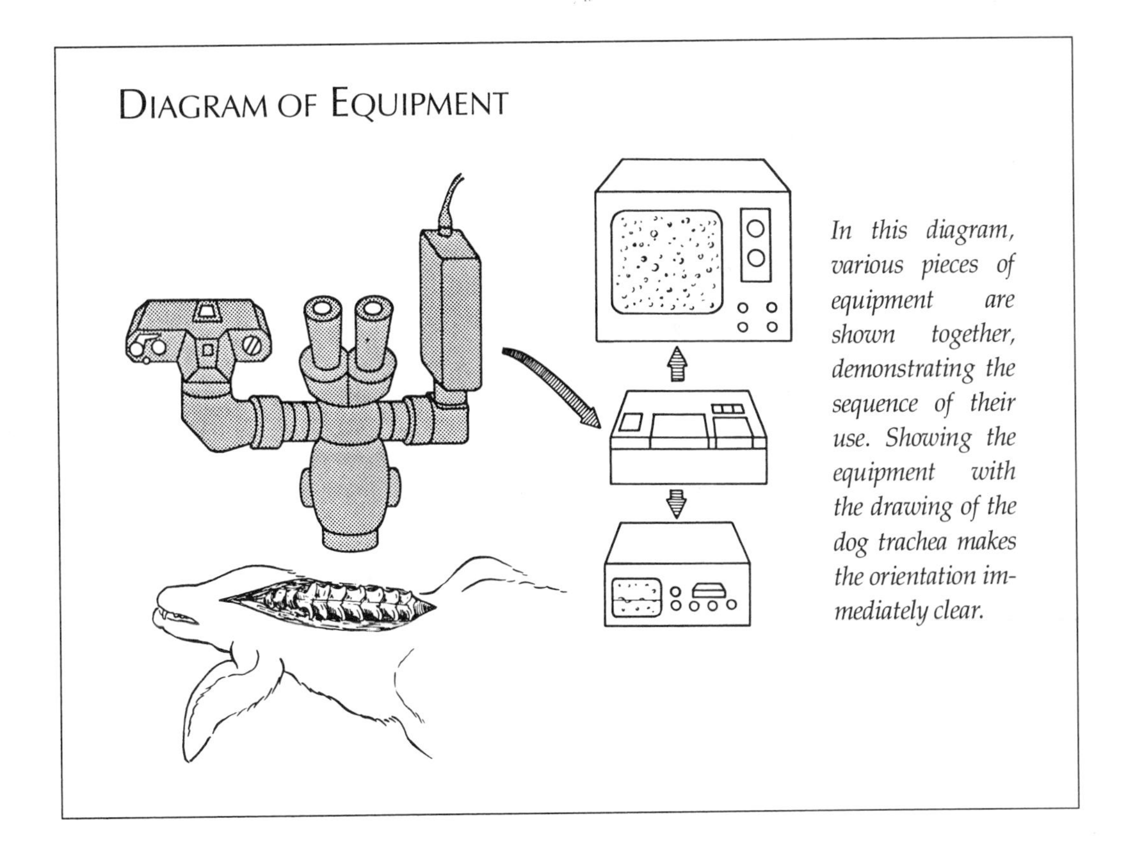

DIAGRAM OF EQUIPMENT

In this diagram, various pieces of equipment are shown together, demonstrating the sequence of their use. Showing the equipment with the drawing of the dog trachea makes the orientation immediately clear.

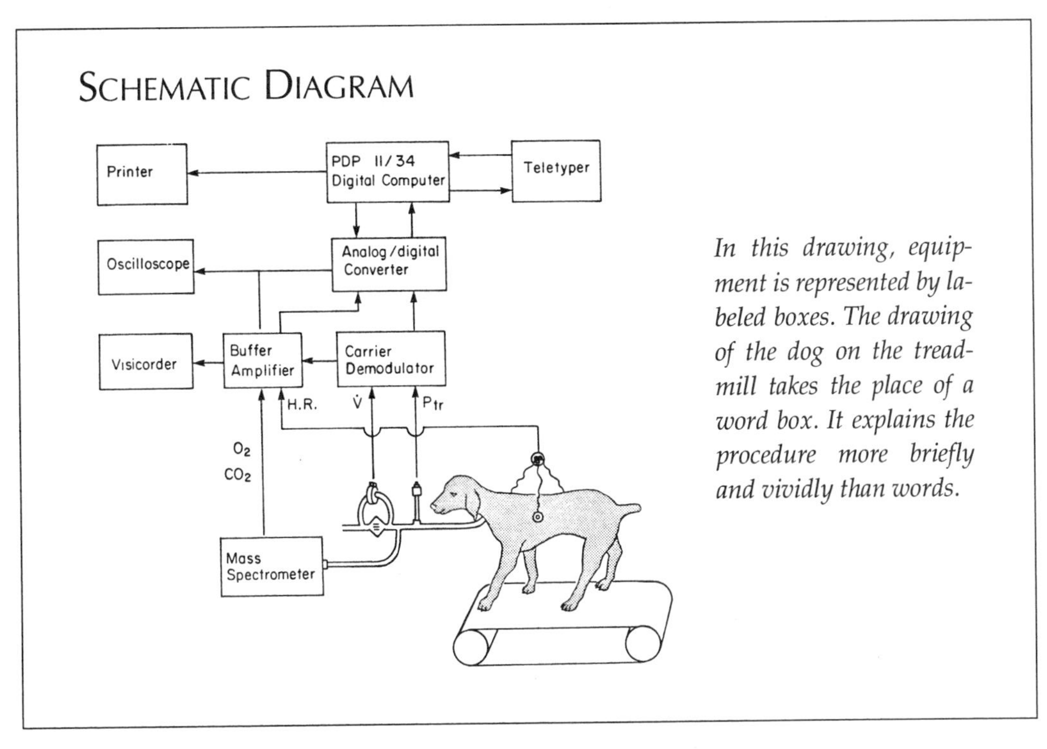

SCHEMATIC DIAGRAM

In this drawing, equipment is represented by labeled boxes. The drawing of the dog on the treadmill takes the place of a word box. It explains the procedure more briefly and vividly than words.

Line or Continuous-Tone Drawing?

The terms "line" (or "high contrast") and "continuous tone" (or "halftone") are used often in photography and printing and are discussed in Chapter 3, "Photographs," and Chapter 7, "The Journal Figure." In the context of the drawings themselves, the terms refer to the materials and techniques used.

A line drawing is done in pen and ink or any other medium that results in black and white only. (Most computer drawings are line drawings.) In a line drawing, tones of gray may be achieved by stippling, fine contour lines, hatching, or cross-hatching. A stipple pattern is made up of small black dots.

A continuous-tone drawing may be done in pencil, watercolor wash, or air-brush. This results in a range of grays. Fine detail and great realism are possible in this technique.

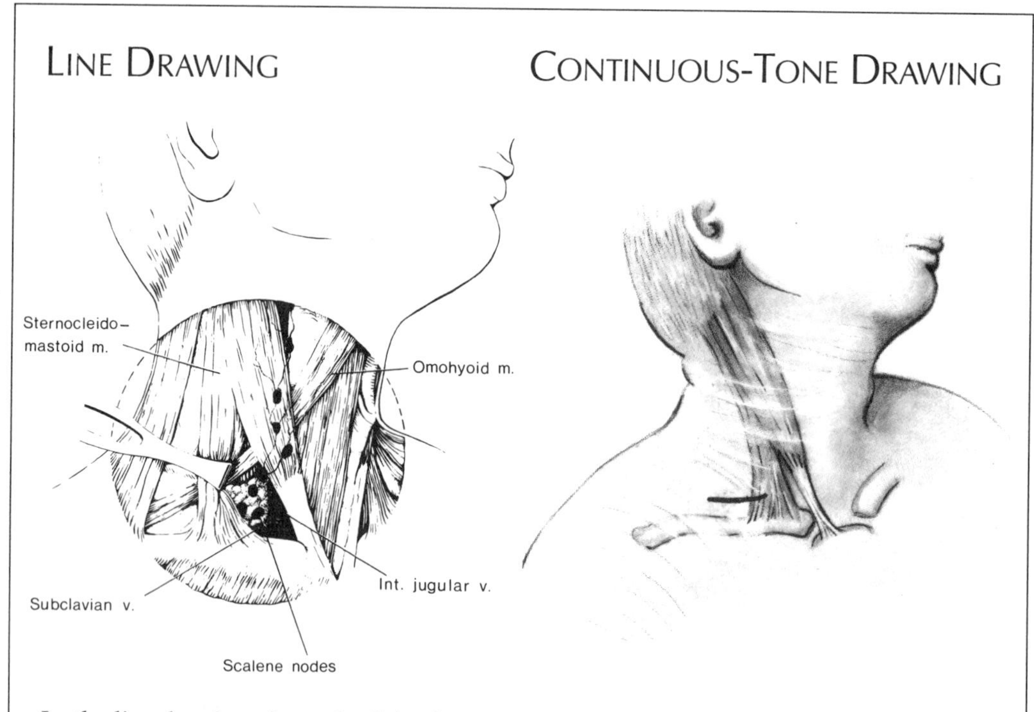

In the line drawing above, detail is shown only by lines and solid black areas. Shading is achieved by lines. The continuous-tone drawing was done with carbon pencil and the "dust" or fine scrapings from the pencil brushed onto the paper to create tone.

Continuous-tone drawings make effective slides because, with enlargement, subtleties and a wide range of tone are preserved. However, for the printing process, the drawings must be converted into an array of tiny dots (halftone dots). Thus, with reduction to one column of a journal, similar to the reproduction on the preceding page, some range of tone disappears, and the darkest areas might fuse together and become black.

Fine lines may be lost in a line drawing as well when reduction is great. Care should be taken to be sure lines are sharp and black. Be sure that the dots of any stipple used are large enough not to fade away and well enough spaced not to fuse or turn to black in reduction.

DOING DRAWINGS

Not all drawings are done by illustrators. Many scientists do diagrams and drawings effectively either by hand or on the computer. Many computer drawing programs are easy to learn and use. Many hand-drawn diagrams and drawings are powerful communicators.

Whoever will do the drawing, keep in mind

- The purpose of the drawing, the important points of the information.
- The medium to be used—slide, paper, or poster.
- The audience to be addressed.

If using an artist, make the project a collaborative effort between author and artist. (See Chapter 10, "Using an Illustrator.") If drawing by hand or computer, expect to spend more time than you think. Ideas should be roughed out, maybe more than once. Materials and drawing programs should be mastered. (See Chapter 11, "Using a Computer," and Chapter 12, "Drawing by Hand.")

Although drawings are challenging and usually time-consuming, an appropriate drawing can communicate faster and more dynamically than words. It can clarify, simplify, and emphasize information in an interesting and inspiring way. Drawings are a priceless help in communicating information.

3

PHOTOGRAPHS

Photographs may be essential for validation and documentation. X-rays, photomicrographs, and photographs of patients and specimens are examples.

Photographs are also essential for reproducing figures you have produced.

In illustrating scientific information, photography will be used in two different ways:

1. Reproduction of figures for slide, publication, or poster of the illustration that has been produced by drawing or photography.
2. Production of a figure by photographic means.

Although this chapter concentrates on the latter, production of a figure by photographic means, it is important to recognize or review the requirements for good reproduction.

PHOTOGRAPHIC REPRODUCTION

Good photographic reproduction of drawings or photographic figures can mean the difference between a figure that is clear and legible and one that frustrates because it cannot be seen easily. Responsibility for good reproduction rests with both the author and photographer. For best results, both the author and the photographer should cooperate and be clear about their requirements.

Following are listed some of the requirements of both author and photographer. Keep these in mind while still in the planning stage. It will ultimately save time and money and result in photographs that will enhance the project.

An author requires of the photographer:

- Fidelity to the original. Any enhancement or manipulation must not distort the original.
- Proper enlargement or reduction.
- Good resolution.
- A full tonal range for continuous-tone subjects.
- Sharp contrast for black and white subjects.

A photographer requires of the author:

- Clear, clean originals.
- Complete, fully labeled originals with complete arrangement of all separate parts.
- Uniform density of line in the original work.
- Uniform background for color or continuous-tone originals, without paste-ups, cuts, or white-outs.
- Specific and clearly stated directions as to reduction, contrast, grouping of pictures, and emphasis.
- A format that is appropriate for the medium intended.

For best results, consult the photographer while in the planning stage. Explain to him or her the medium that will be used and the purpose of the figure. A photographer can suggest the size that is most convenient to work with, the format and dimensions of the paper and film being used, and whether a figure may be effectively enlarged or reduced.

In working with a commercial photographer, discussion in the planning stage covering questions, needs, and expectations is especially important. Show the photographer examples of the kind of reproduction you want. If you want to do your own photography, consultation in the planning stage with a professional photographer may be invaluable. To find a photographer who specializes in scientific and medical photography, contact a nearby medical school or hospital.

Tracings

Examples of physiologic tracings illustrate some of the problems in reproducing figures and point out the importance of interaction between photographer and author in solving those problems.

Shown in this section are tracings that are recorded on paper. However, many laboratories have recording equipment connected directly to a computer program. These data can sometimes be transferred to a drawing program in which the choices and suggestions mentioned here may be made easily.

Below are electroencephalographic (EEG) tracings that record sleep patterns. They are recorded on paper with a background grid, so that the first decision to make is whether to keep the grid. It is possible to eliminate colored grids photographically by using filters. If both grid and tracing are the same color, for example, black or red, it is impossible to eliminate the grid. A photographer can easily determine whether the grid can be eliminated.

Sometimes tracings are faint and uneven. A photograph or sometimes even a photocopy of the tracing from a copier may darken faint lines and increase the contrast. Try the photocopy machine before consulting the photographer.

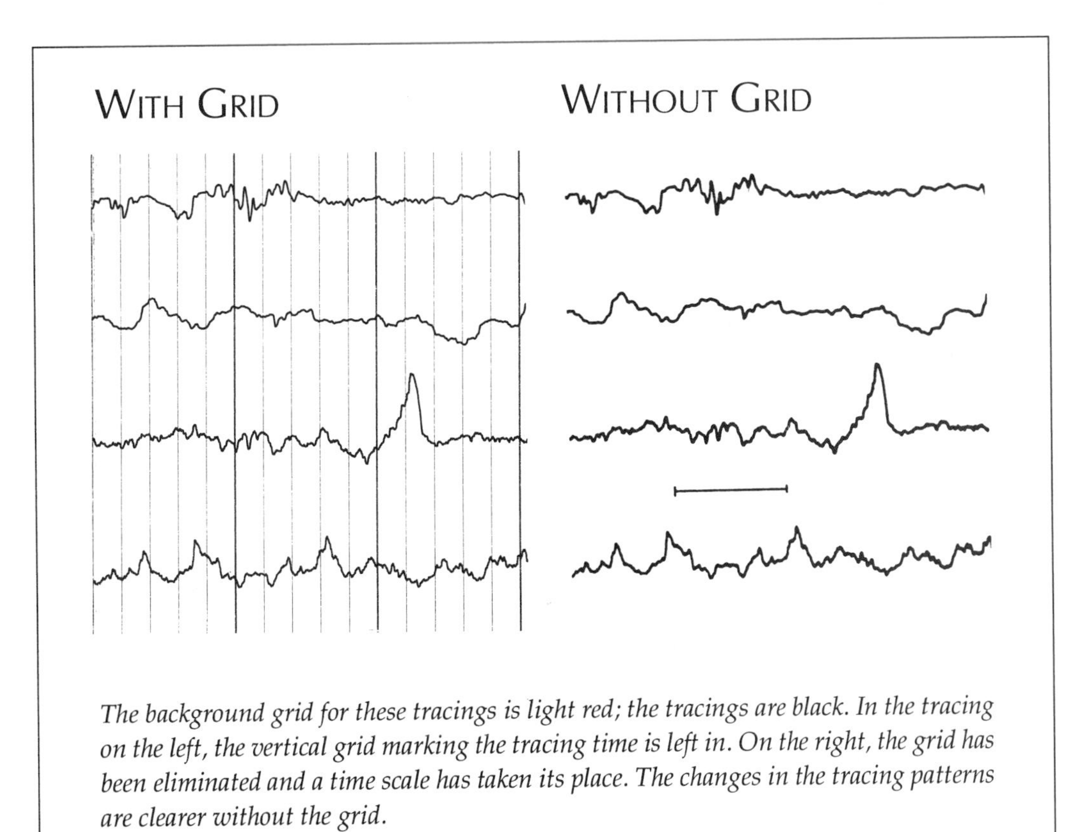

The background grid for these tracings is light red; the tracings are black. In the tracing on the left, the vertical grid marking the tracing time is left in. On the right, the grid has been eliminated and a time scale has taken its place. The changes in the tracing patterns are clearer without the grid.

The best photographer, however, cannot separate overlapping tracings.

Settings for tracings should be planned with possible reproduction in mind. Remember this also when marking tracings. Do not expect the photographer to

OVERLAPPING TRACINGS

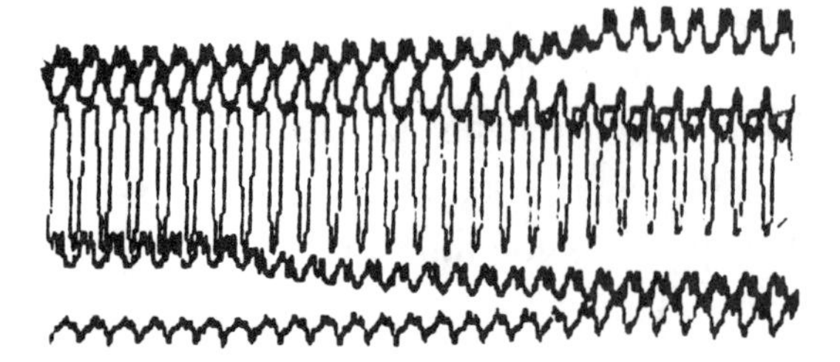

These tracings were set too close together. Although an artist could separate them by laboriously redrawing each curve, it would be more efficient to redo the tracing using appropriate settings to separate them.

touch up or erase thick indelible notes made on the part of the tracing to be used. To save time, and for better final results, mark the tracing with pencil or mark it above or below the part that will be used.

Selection and mounting of tracings that will be used together are the author's responsibility.

SELECTION AND SEPARATION

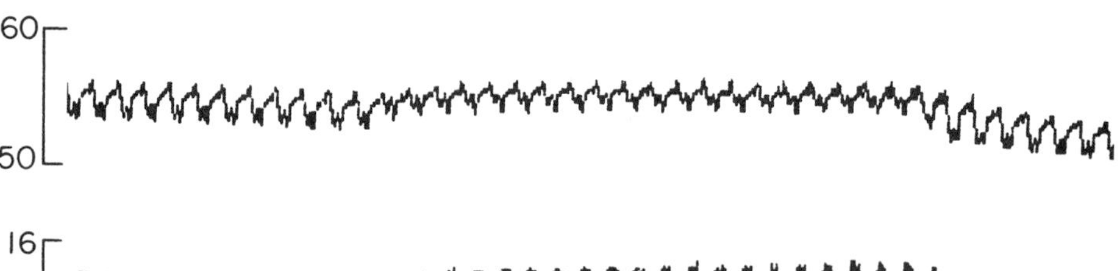

These tracings were parts of larger tracings, selected for their relevance. The light computer print-out was photocopied to darken the lines. Because the left-hand scales overlapped, they were cut apart and mounted with space between them. Scales were redrawn and numbers added by hand.

Use only the parts of the tracing pertaining to the point to be made, keeping in mind the final shape and reduction of the figure. A photographer can assess how effectively the lines will reduce and can advise about format for the medium to be used.

Often many tracings are shown together. Extraneous parts of the tracings must be eliminated and relevant tracings should be placed in a logical order.

Repetitious labels should be eliminated and labels added that will fully clarify your information.

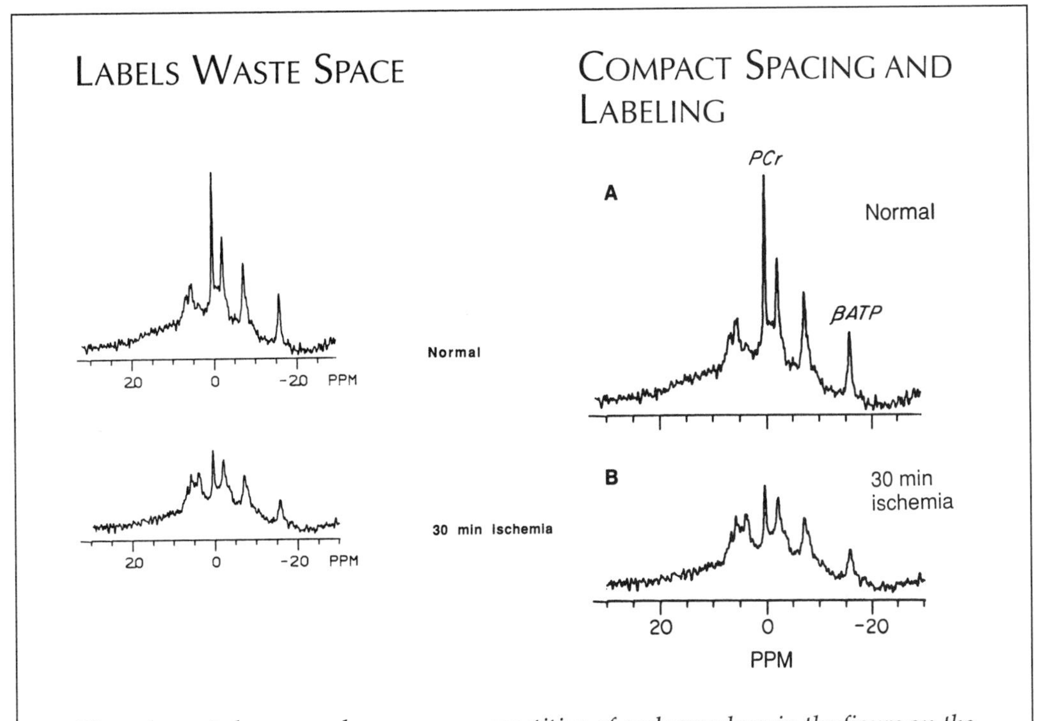

There is wasted space and unnecessary repetition of scale numbers in the figure on the left. On the right, repetitious numbers were cut, allowing the tracings to be moved closer together. The more compact arrangement of the labels works well for reduction to one-column journal width. Notice the greater reduction of the figure on the left because of its width. The addition of peak labels and the distinguishing letters, A and B, makes the figure clearer to the viewer and easier to discuss in publication or slide.

Labeling of tracings should be complete and appropriate before submission to a photographer. The style of the labels should be thin-line. Bold or thick labels compete with, and usually overpower, the thin tracing lines. Discuss with the photographer whether it would be better to label a photograph of the tracing or the original tracing.

PHOTOGRAPHIC FIGURES

X-rays, photomicrographs, and gels function as data. Photographs from the cathode ray tube (CRT) screens are considered documentation. These photographs give factual information, verisimilitude, or proof.

In addition, photographs may be required of patients or experimental animals for utmost realism. With patients, it is essential to get their permission and preserve as much of their privacy and anonymity as possible. Although it is not necessary to get permission to use photographs of animals, given the controversial nature of working with animals today, it is wise to avoid photographs showing anything that might be construed as cruel. A selected part of the experimental procedure or a diagrammatic drawing might be a better alternative.

Because "reality" and "truth" are essential in these figures, it is important to be straightforward and thoughtful in the selection of the areas to be used. Manipulation such as enlargement, reduction, and increase or decrease of contrast must not distort or change the information. Touch-up is permissible only to eliminate distracting artifacts. Labels should be used judiciously and sparingly, and should not hide or distract from important information.

With the widespread availability of scanners and photo design software such as Adobe Photoshop™, the possibilities for undetectable distortions and changes are so great that special care must be taken not to mislead, to be scrupulously honest.

In this chapter, examples of photomicrographs, gels, and video images are used to show that clear communication of the information in these photographs requires extensive planning and careful treatment.

Plan Photographs Carefully

- Locate the area of interest near the center of the picture.
- Trim unnecessary parts.
- Keep the photograph as close to the final reproduction size and/or format as possible.
- Use as few intermediate steps as possible to go from original to reproduction.
- Use thought and care in mounting photographs together.
- Use labels that are proportionate to the size of the photograph.
- Make labels consistent in size and style.

High-Contrast and Continuous-Tone Photographs

High-contrast film captures only white and black. It does not record gray tones. Continuous-tone film records a full range of tones from white to black.

HIGH-CONTRAST CONTINUOUS-TONE

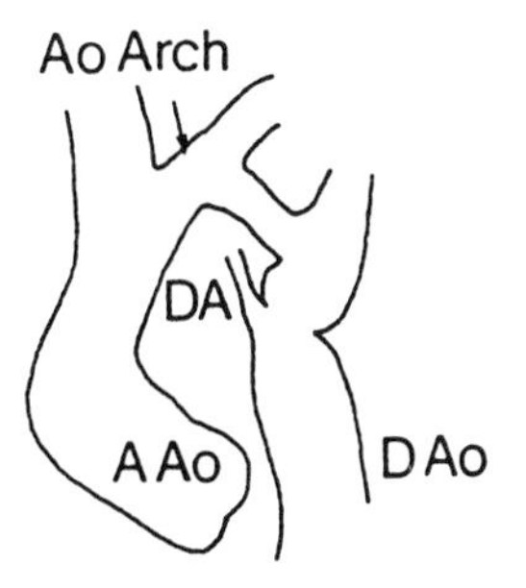

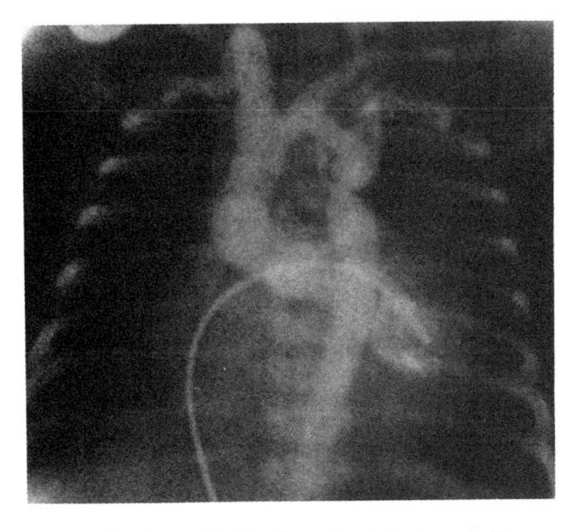

Because the pen and ink drawing on the left includes only black and white tones, it was shot using high-contrast film. The angiogram on the right includes a range of grays and was photographed with continuous-tone film.

The advantage of shooting with high-contrast film is that defects in the original such as white-outs and cut-and-paste lines as in paste-on labels will not show in the final print. Greater care must be taken with the continuous-tone original, and labeling is best done with a transfer type.

PHOTOMICROGRAPHS

Photomicrographs encompass a full tonal range from white to black and are shot with a continuous-tone film. Fine, crisp resolution of detail is the goal of these photographs. However, careful planning is the essence of their final presentation.

Size of Prints

Before the photomicrograph is printed, know what the final size will be. This is particularly true for journal presentation. There should be no reduction from the print to final reproduction in a journal. Plan the dimensions to fill the column or page to avoid wasting space. Plan whether to group images into one figure. Plan the final size to illustrate the main point. Plan whether you will cover your image with labels or show a labeled diagram next to the image.

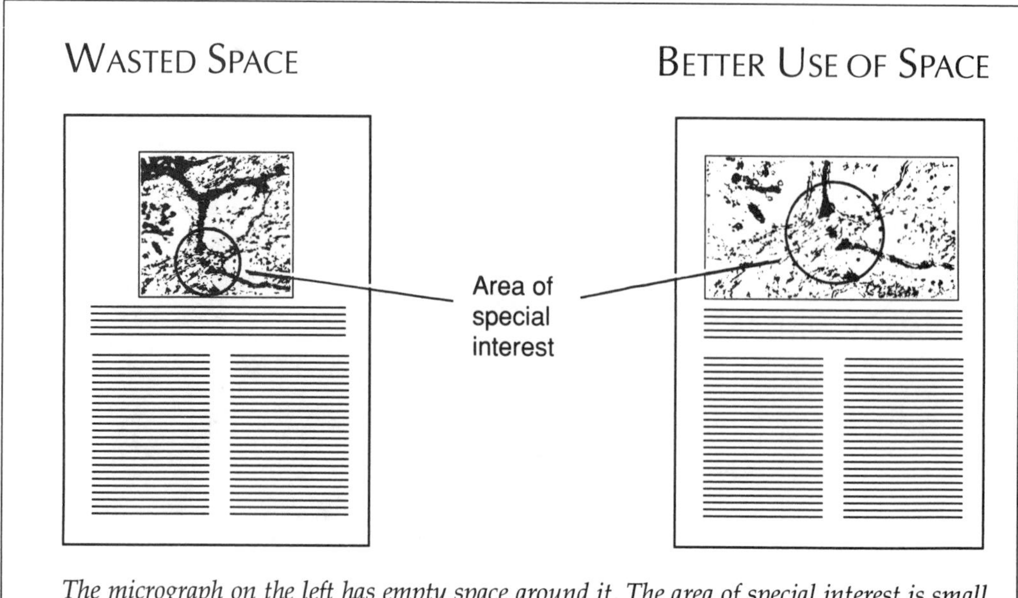

The micrograph on the left has empty space around it. The area of special interest is small and at the bottom of the figure. The figure on the right fills the page space. The area of special interest is centered and enlarged.

Templates for Page Dimensions

Before making final prints, draw a template duplicating the exact page dimensions. Done quickly and in pencil, this will indicate the shape of the final print, how it will fit in the page, and where the caption will be positioned (see opposite page).

Use a template to determine the exact dimensions of the final figure before the print is made. This is especially important when arranging groups of figures. Do not expect the journal to group figures together. This requires time, thought, and experimentation and should be done before making final prints. Determine the focus and magnification for each separate figure on the basis of their relationship to each other. The weighing of priorities, in turn, may be influenced by the space the photographs will occupy.

Arrangement

The grouping of figures should be logical according to related content. The images to be grouped should not be widely different in tonality. The composition of the grouping should be balanced. Decide ahead of time whether the final figure will be in a one-column space, half page, or a full page. If it is to be a slide, decide ahead of time which slide format will be used. The space between the separate figures should

PAGE AND FIGURE TEMPLATES

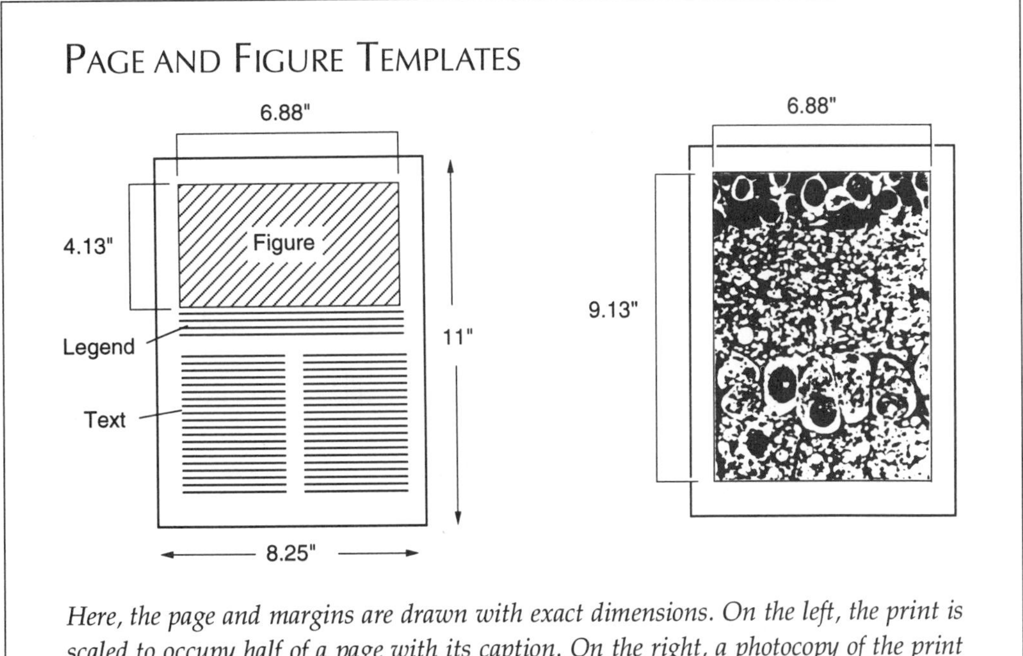

Here, the page and margins are drawn with exact dimensions. On the left, the print is scaled to occupy half of a page with its caption. On the right, a photocopy of the print occupies the whole page without its caption. It is acceptable to put the figure caption at the bottom of the facing page.

not be so great that it will waste space or appear as a glaring white grid pattern between photographs. However, the space should be large enough to show the figures as clearly separate (see figures B and C on the next page).

The arrangement of three figures together poses special problems, as shown on the next page and page 27 (top). If all three are of equal significance, they should be the same size and placed side by side as in figure C on the next page. If one is clearly more important, it might benefit by expansion and enlargement as in figure B.

If one figure is clearly subordinate to the other two, consider whether the least important figure could be inset into the figure with which it is most closely related, as in the bottom figure on the next page.

For slides, a horizontally rectangular or square arrangement works best. Many speakers do not label micrographs. Labeling of even large negatives is difficult and not really necessary, because the lecturer can point to and describe the significant areas.

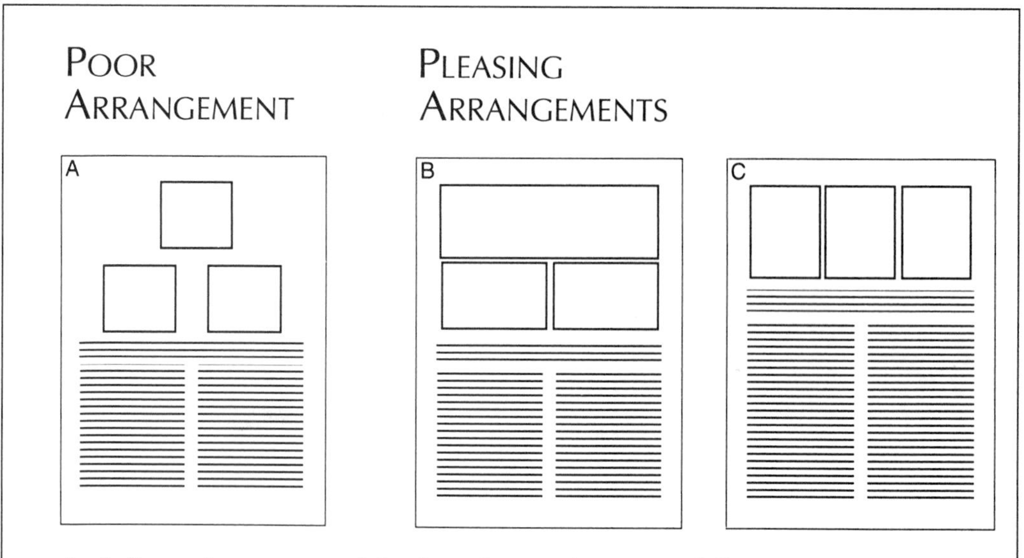

In A, the empty space around the three photographs gapes and distracts. In a slide, this would be especially pronounced. The relationship between the three photographs is not clear. In the two arrangements on the right, the figures become an integral part of the page, and the eye is led naturally and easily from the photographs to the caption, and on to the text.

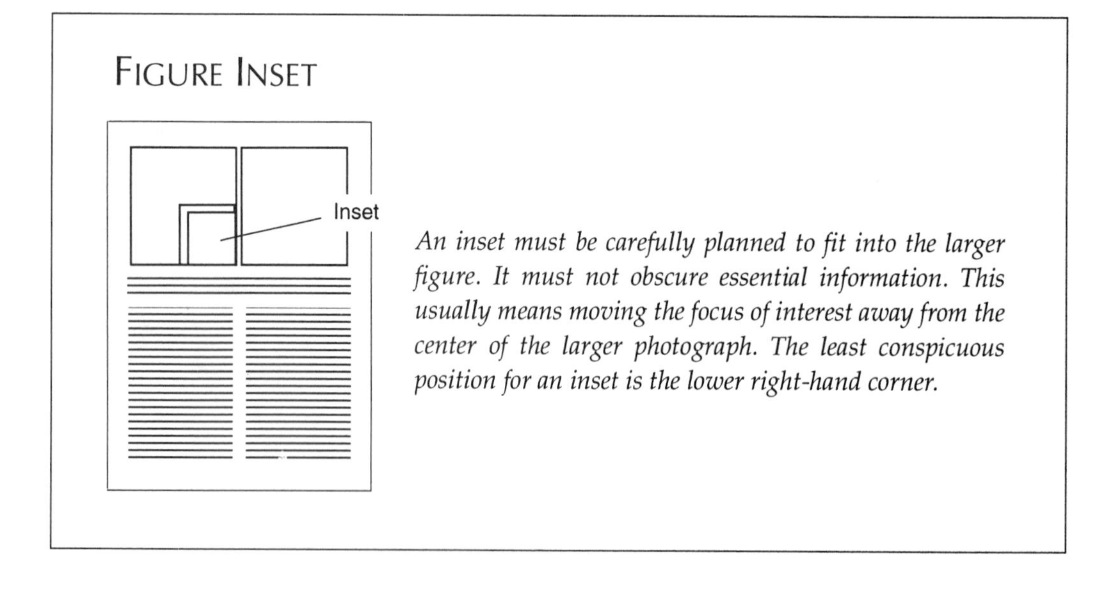

An inset must be carefully planned to fit into the larger figure. It must not obscure essential information. This usually means moving the focus of interest away from the center of the larger photograph. The least conspicuous position for an inset is the lower right-hand corner.

VERTICAL ARRANGEMENT

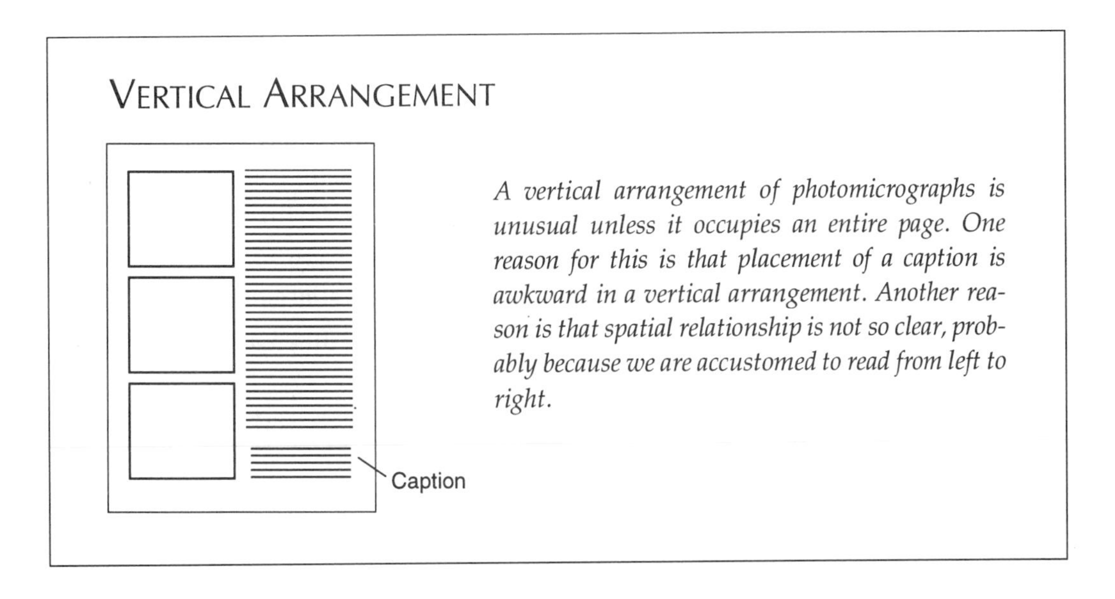

A vertical arrangement of photomicrographs is unusual unless it occupies an entire page. One reason for this is that placement of a caption is awkward in a vertical arrangement. Another reason is that spatial relationship is not so clear, probably because we are accustomed to read from left to right.

Labels

Labels are secondary to the picture and should be short. They should serve as cues to the explanations in the caption. Labeling should be done directly on the photograph using white or black transfer or rub-on letters. These may be bought at art

TOO LARGE

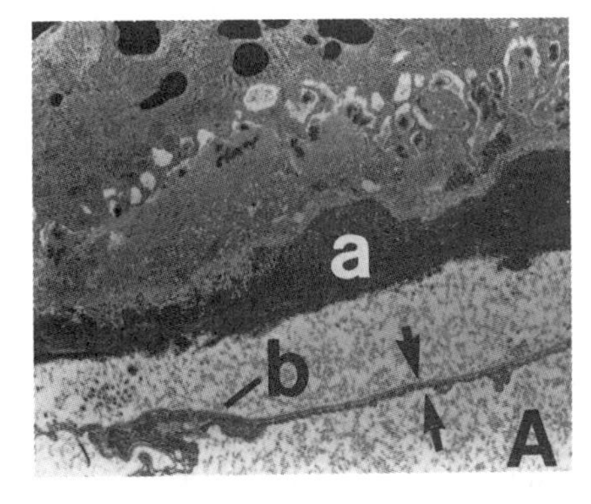

WELL PROPORTIONED

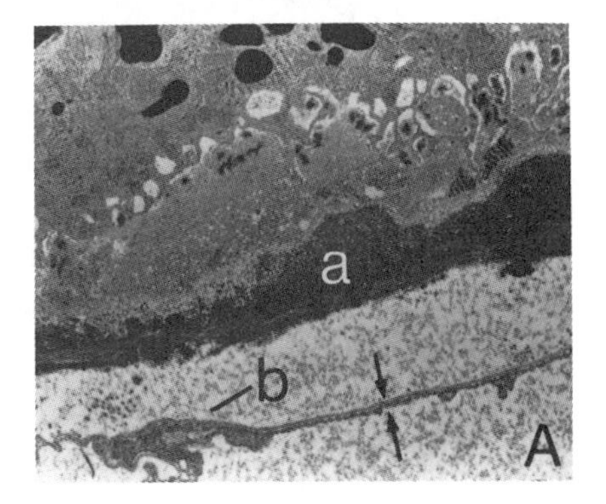

The labels on the left are 24-point Helvetica font in a medium thickness. This size clamors for attention. It is the first thing seen. On the right, the labels are 18-point Helvetica light. They are an integral part of the photomicrograph and inform without distracting. Notice that the lower case of the same point size font is smaller than the upper case.

supply stores or some stationery stores. Choose a simple font that will not distract. Choose a size that will be legible, and use it consistently. If all your figures are planned for no reduction to final size, the same size labels may be used throughout. Because journal text size is usually 10 point, an 18 point-label works well. If in doubt as to the best size, photocopy the image and label the photocopy before the final labeling.

One problem with using rub-on labels is that it is easy to run out of the numbers or letters that are frequently used. Be sure to keep a large supply on hand. Another reason for standardizing to one size is to avoid the temptation to use another size in place of the most appropriate size.

Different placement of the labels may be tried first on photocopies. Do not obscure important details with labels. If many labels are essential, consider making a diagram of the image, which can then be labeled without cluttering or covering the image.

LABELING BY DIAGRAM

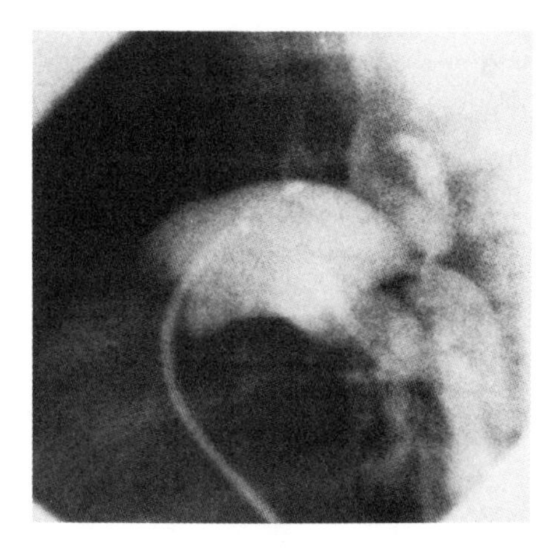

The diagram labels serve to keep the photograph's details clear. The diagram also clarifies structures in the photograph. Both are mounted together to be used as one figure.

A problem in labeling photomicrographs and other continuous-tone pictures is whether to use white or black labels. There are times when neither works well. On the next page are some solutions.

For labels longer than one letter or number but limited to short words or few letters, rub-on labels can be applied to gummed circles or ellipses that may be fixed to the photo. Gummed circles and ellipses come in different sizes and may be bought in stationery stores.

WHITE BACKGROUND LABEL

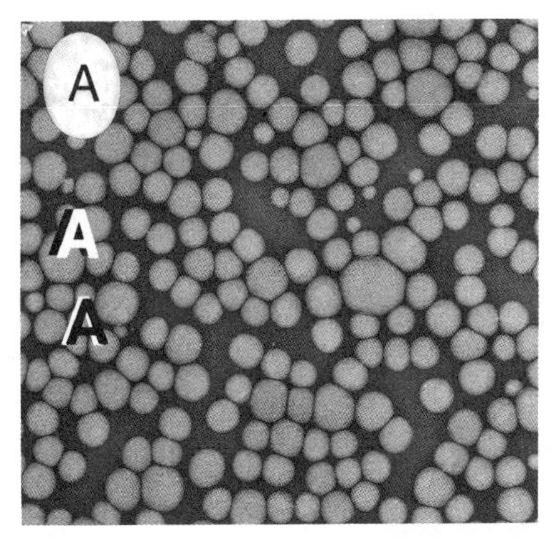

The top label with the white background is very clear. This style label is available in rub-on (transfer) letters as single letters or numbers. The white label in the middle outlined with black is less easy to see. But the black label at the bottom, despite being outlined by white and black, disappears into the background.

Arrows and lines are available that are both black and white. These may work best in difficult areas.

Consider selecting and cropping prints in such a way that there is a space that is either dark or light enough for labels to show up.

LINES AND ARROWS

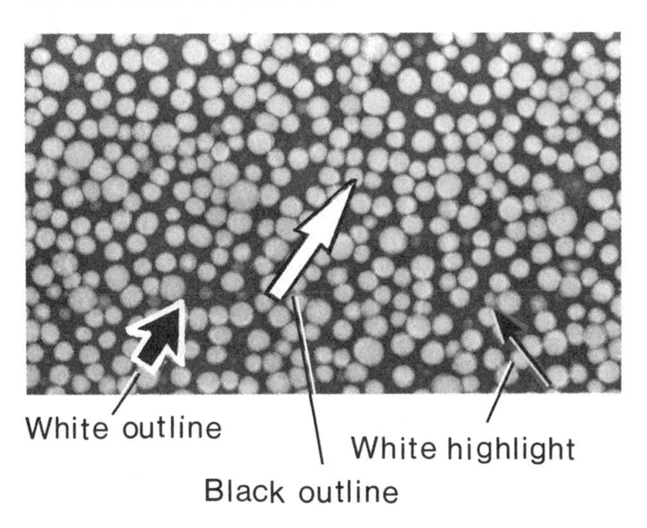

The overall tonality of this image is neither dark nor light, so that neither black nor white labels show up well. Black rub-on labels are highlighted only on one side. The black arrow outlined thickly with white shows up best. Several styles of arrows are helpful to point out different areas.

SCALE MARKING

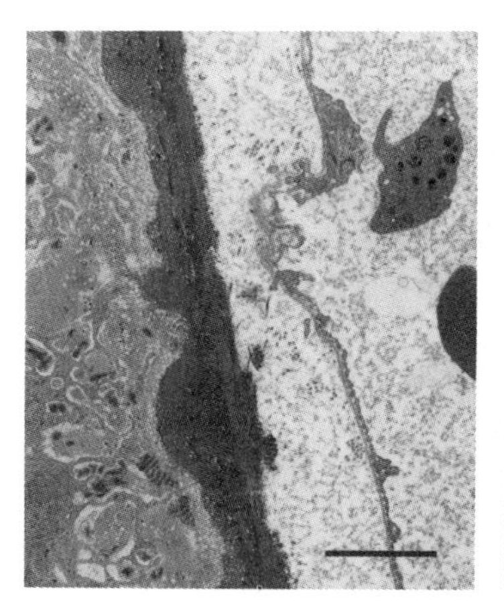

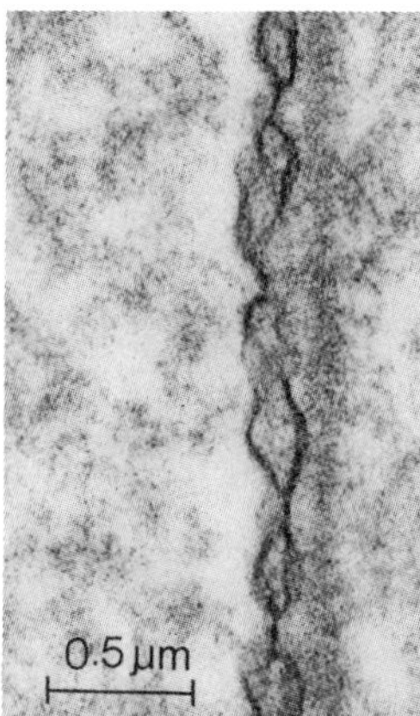

A scale indicating magnification should be shown on the photomicrograph. A single thick line to the exact dimensions (left) is acceptable. The scale units themselves are then indicated in the figure caption. Or dimensions may be labeled directly on the figure (right). Here the thin scale with end ticks works well against a light background.

GELS

Gel patterns show the array of nucleic acids and proteins separated by electrophoresis into bands of different components, different molecular weights, or restriction fragments. Electrophoresis separates serum proteins into identifiable components. After separation, nucleic acids and proteins may be stained and quantified.

Gel electrophoresis is a powerful analytical tool and is widely used in genetic and peptide mapping. The photography and labeling of gels require special consideration. A light background is desirable, with contrast between the darker and lighter bands. Labeling should be legible but unobtrusive.

A problem in the reproduction of gels is that there may be a loss of quality (degradation) with each photographic step away from the original. Labeling must be done on a photographic print of the gel. A negative and print of this print must be made to satisfy most journal requirements for several glossy prints. Then the printer makes another negative and print for publication. Thus, it is most important to have a crisp, sharp first print.

It is possible to scan the photograph of the gel into the computer, trim, label, enlarge or reduce it in a program such as Adobe Photoshop™, and then print it directly from the computer. The quality of the print, however, depends on the printer resolution, and subtleties of the gel may be lost in this method.

UNLABELED GEL

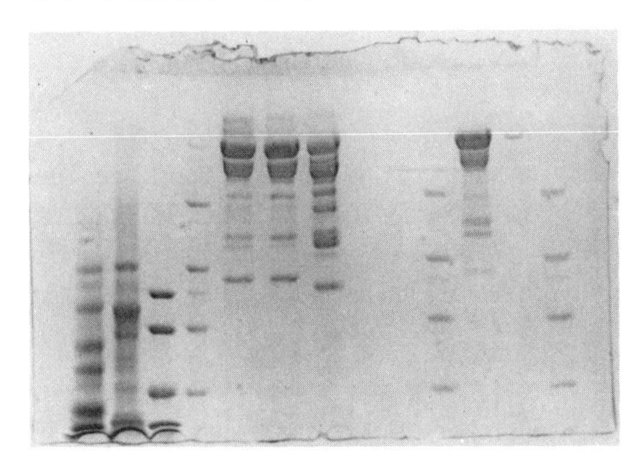

Molecular weight of proteins is assessed in this polyacrylamide gel electrophoresis using a sodium dodecylsulfate buffer (SDS-PAGE). The wet gel was photographed on a light box using a fine-grain film. Good contrast was achieved by experimentally adjusting exposure and time and by manipulation in printing. This gel was made by Geri Chen and photographed by David Hardman for the Cardiovascular Research Institute, University of California, San Francisco.

Although this method is not the best for final prints, scanning the gel at 100% with no change in size is useful for labeling the gel. In the scanner software or a photodesign program, the gel can be trimmed or rearranged, bands can be labeled, and the result can be printed. The photograph of the gel can be pasted over the print for accurate labeling that is easy to do and to change.

It is advisable to send the original labeled print (or the disk if done on the computer) to the journal. The journal can then make a negative and print from the original. Some journals require original work, others photographic prints. Not all journals will accept computer prints so it is important to check journal instructions.

Labels will not always fit on the photographic print of the gel. The white surrounding area of the print is usually not wide enough for labels, and labels on the print itself may cover important information. In this case, trim and mount the gel print on white backing paper or board. Trimming and mounting may be done at this time to rearrange, add, or eliminate gel sections.

In the original labeled print, the labels are high contrast (black and white only) and the gel is continuous tone (containing shades of gray). However, in the reproduction shown in the top figure on the next page, everything is continuous tone; the background to the labeling is gray. If separate negatives are made of the gel and the surrounding labels, the result is more like the original. This is more time-consuming for the photographer, but it results in a clearer, crisper final reproduction (see the bottom figure on the following page).

LABELED GEL

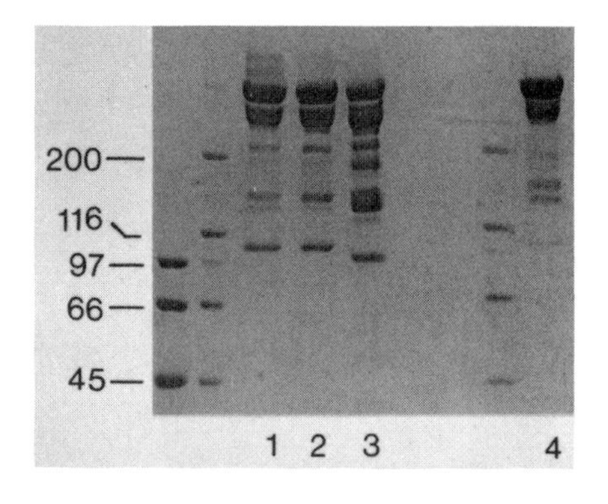

The labels on this gel are 10-point size. The style is simple, sans serif, and thin-line. It is legible but does not distract from the subtle tones of the gel bands.

GEL PRINTED FROM TWO NEGATIVES

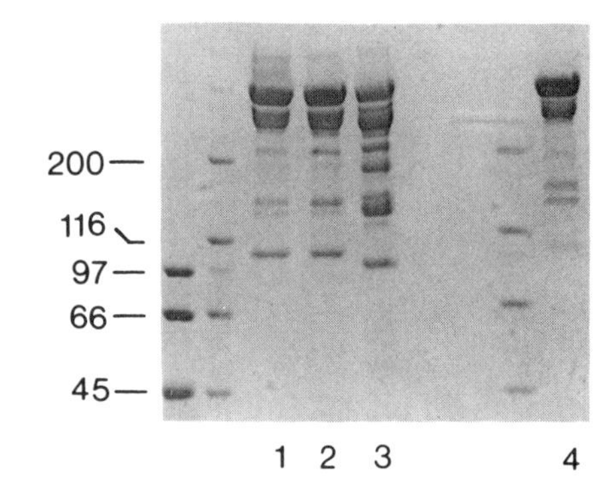

Printing of this figure was done from two negatives. Notice that the background for the labels is pure white, whereas the gel itself contains a range of grays.

The same standards for label size and thickness apply also to immunoblots. For immunoblots, the gel is transferred onto a nitrocellulose filter, where it is subjected to a complex of protein antigens and antibodies. It is then exposed to X-rays or to a dye. A print may be made of this and then labeled, or better yet, labeling may be done directly on the film. If trimming and rearrangement of the columns is necessary, the film may be cut and taped to a clear acetate backing, which may then be labeled.

In using transfer labels, stay away from decorative or complicated script types. Choose instead a plain, sans serif type face such as Helvetica. Make sure that you have enough of the correctly sized letters so that you will not have to use an inappropriate size or font.

VIDEO IMAGES

Photographs are often taken from video screens. Echocardiography, ultrasound images, computed tomography (CT), scintigraphy, nuclear magnetic resonance imaging (MRI), and others are photographed from the video screen. Their effectiveness depends on the resolution of the screen, the ability to enlarge crucial areas, and the labeling options that the computer program offers.

Below is an example of ultrasound imaging as it comes from the computer screen.

UNIMPROVED

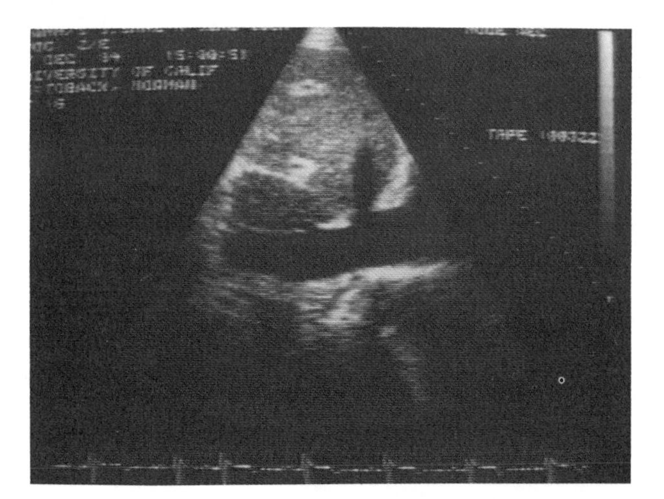

This is an echocardiogram showing sections of the inferior vena cava and the hepatic vein. The background labeling (illegible at this reduction) includes the mode, tape number, place, patient name, and date. At the bottom is an electrocardiogram (ECG) tracing. Althgough this is vital information to the researcher, it is usually irrelevant to the audience and should be left out.

Because most equipment does not allow for changes on the screen, cropping and labeling must be done on the print.

IMPROVED

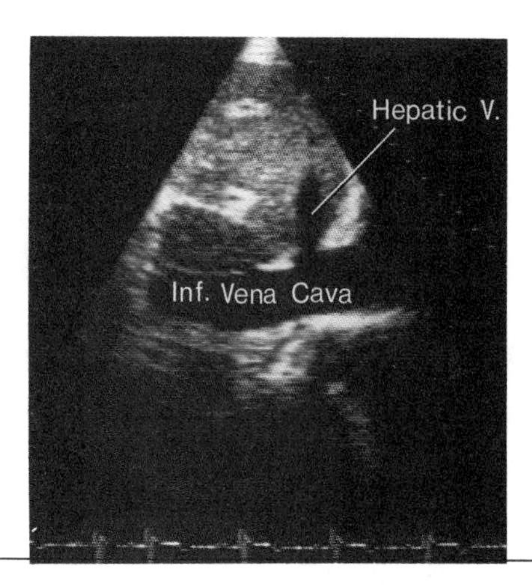

This is the same image. It has been cropped on the sides, and the background labels have been blocked. White transfer labels have been applied directly to the print. The electrocardiogram tracing has been removed.

REMOVE THE BACKGROUND

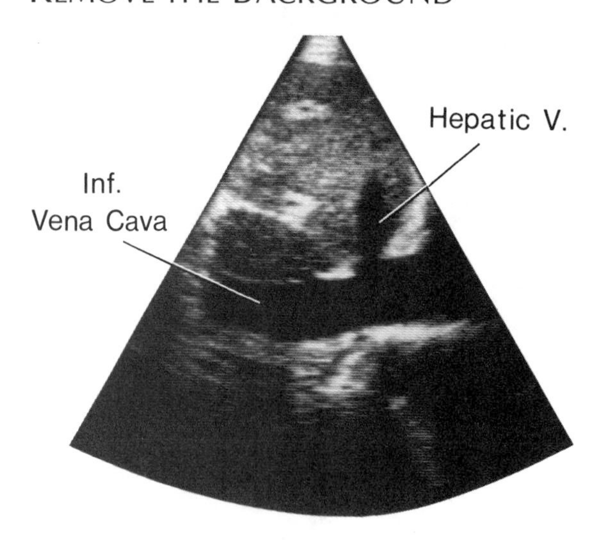

The fan-shaped echocardiogram was cut from its black background as an easy way of eliminating unwanted background labeling and lines. The black labels on a white background are crisp and clean.

An echocardiogram has a fan shape, which is lost in the black background of the figures on this and the previous page. With the background removed, black labels can be added to the surrounding white area, giving an uncluttered look.

If it is not possible to program the equipment to eliminate or enlarge labels or to enlarge certain areas and eliminate others, it must be done by hand. Although it takes time to do this, it will make the difference between communicating or confusing. In Chapter 12, "Drawing By Hand," simple processes for trimming, mounting, and labeling tracings and gels are described.

Digital acquisition equipment is available for images such as CT scans, MRI, echocardiograms, ultrasound, and X-rays. Once digitized, the image can be cropped and transferred in a PICT file format to a photodesign program. In such an application, label size, addition, or removal is easily accomplished. The digitized image is stored on an optical disk.

For example, Peter Callen of the University of California San Francisco can transfer a digitized and cropped ultrasound image into the Adobe Photoshop™ application. There he is able to enlarge, reposition, add arrows to, and label the image. Further he can delete areas and change grays, brightness, and contrast.

For slides or prints, he transfers the enhanced image to the PowerPoint® program, where the image can be set into a color background and printed onto film for a slide. The resulting file can also be printed on a color or black and white printer or the file itself can be sent on disk to a journal for printing.

Alternatively, Dr. Callen can scan the ultrasound image on a scanner and transfer it to the Adobe Photoshop™ or PowerPoint® program for slides or prints.

Pictorial archive computer systems (PACS) that digitize images are used in many hospitals as an alternative film storage. These images may be transmitted worldwide.

4 Charts and Tables

Very often the same data can be presented as either a chart or a table. A chart contains pictures, words, or numbers and is outlined or mapped to show sequences, deviations, and pathways. A table is a compilation of numerical or alphabetical data arranged in tabulated and labeled columns showing summaries of findings.

Both charts and tables can be effective vehicles for communication. In fact, they are essential for expressing certain kinds of information. Cycles, algorithms, and genealogies must be shown as charts. Essential numerical values and lists of facts to be compared must be shown in tables.

Charts

A good chart delineates and organizes information. It communicates complex ideas, procedures, and lists of facts by simplifying, grouping, and setting and marking priorities. By spatial organization, it should lead the eye through information smoothly and efficiently.

On the next page are the steps of the nematode life cycle shown in Chapter 1, first listed and then organized into a chart.

LIFE CYCLE LIST

Infective cycle:

1. Infective-stage juveniles
2. Infection of insect larva

Reproductive cycle:

3. Fourth-stage juveniles
4. Adults
5. Eggs
6. First-stage juveniles
7. Second-stage juveniles
8. Third-stage juveniles

LIFE CYCLE CHART

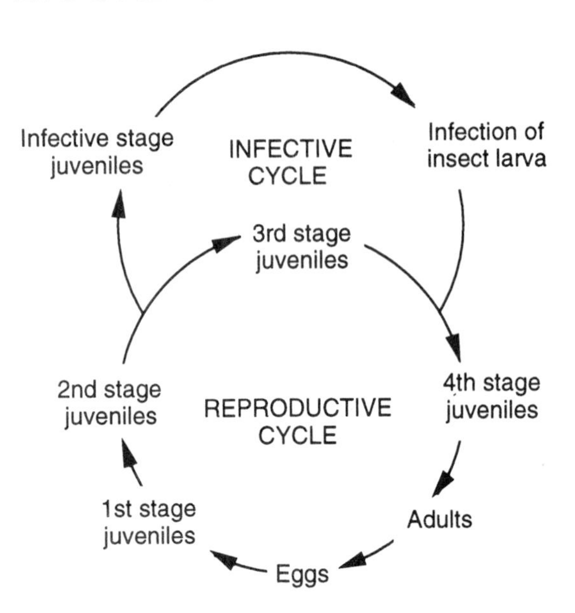

The chart, because of its circular organization, is immediately apparent as a cycle. The relations of the two main divisions, infection and reproduction, are immediately clear. The split path from second-stage juvenile to third or infective stage is clearly shown in the chart. In the same amount of space, the chart is more informative and visually more interesting.

It can be made even more informative and interesting by adding simple drawings.

LIFE CYCLE WITH DRAWINGS

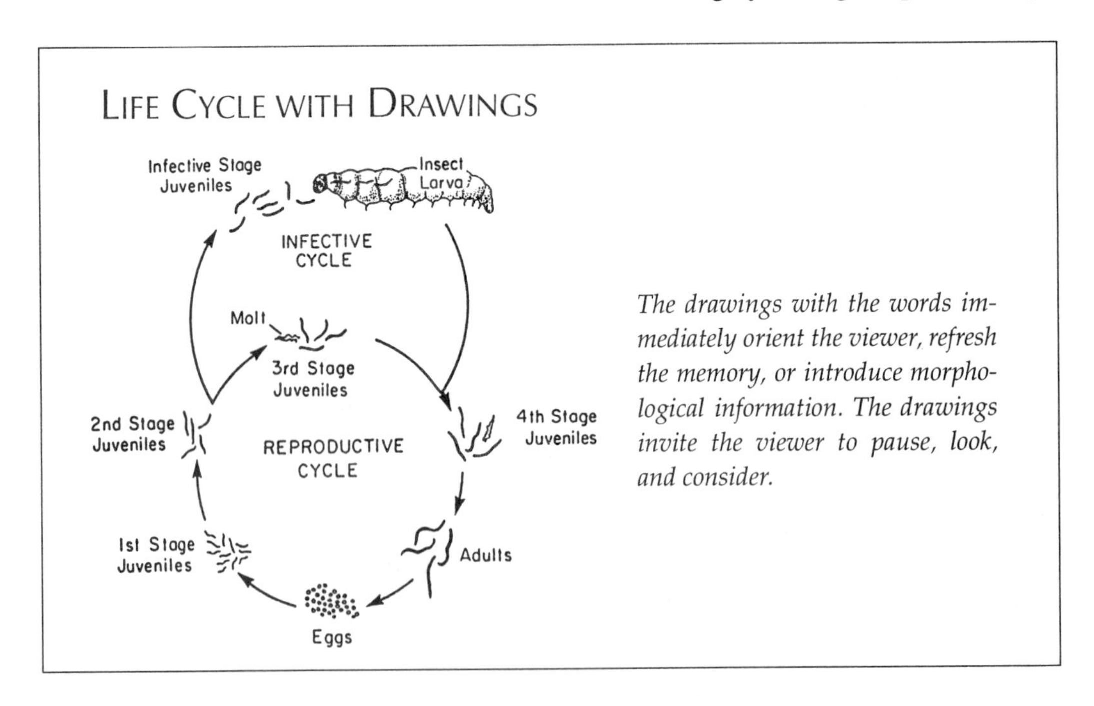

The drawings with the words immediately orient the viewer, refresh the memory, or introduce morphological information. The drawings invite the viewer to pause, look, and consider.

In the following circuit chart of retrograde aortic perfusion, the drawing of the heart actually takes the place of the word "heart." This is possible only if the subject is immediately familiar and recognizable. The boxing in of words sets them apart and suggests the simplified representation of the labeled entities.

CIRCULAR ARRANGEMENT WITH BOXES

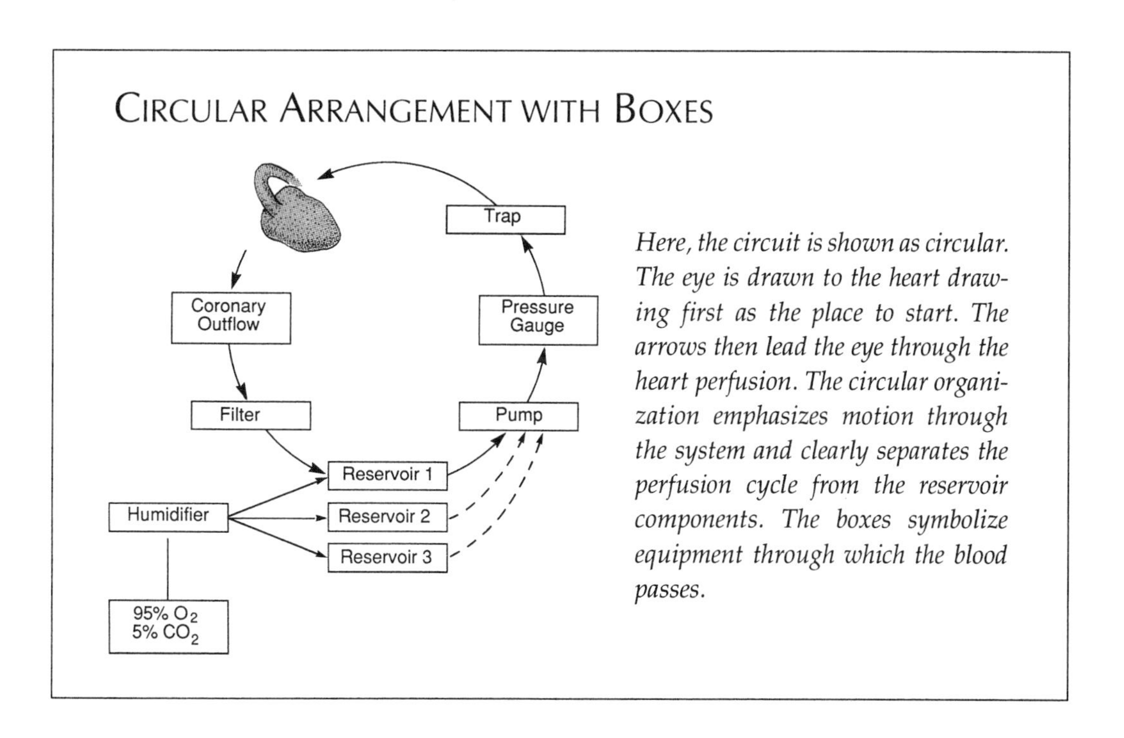

Here, the circuit is shown as circular. The eye is drawn to the heart drawing first as the place to start. The arrows then lead the eye through the heart perfusion. The circular organization emphasizes motion through the system and clearly separates the perfusion cycle from the reservoir components. The boxes symbolize equipment through which the blood passes.

RECTANGULAR ARRANGEMENT

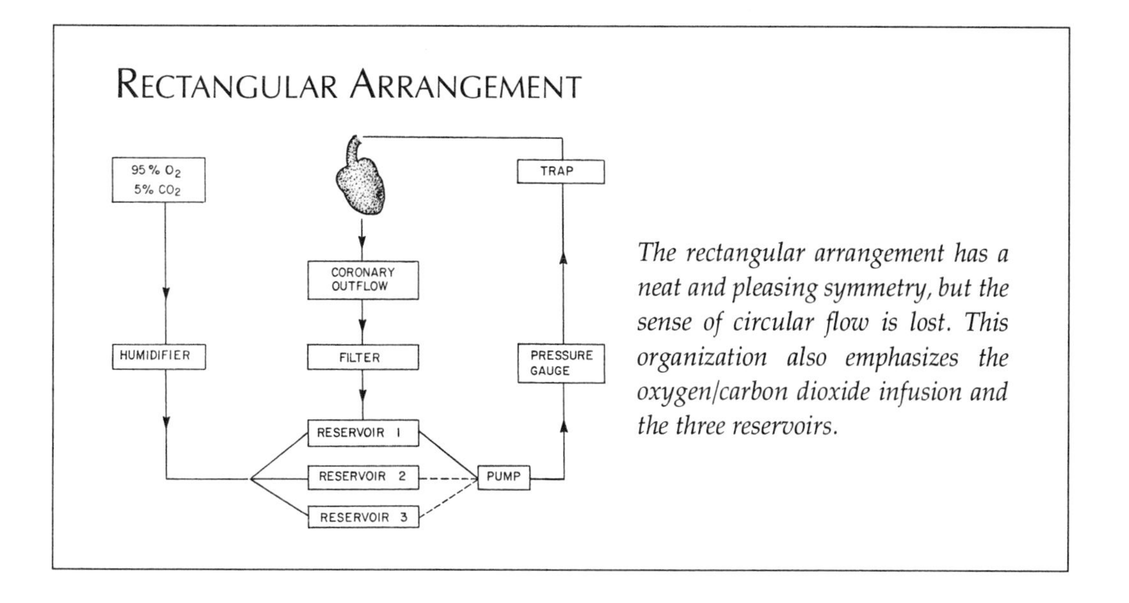

The rectangular arrangement has a neat and pleasing symmetry, but the sense of circular flow is lost. This organization also emphasizes the oxygen/carbon dioxide infusion and the three reservoirs.

A schema presents constituents in a pattern or plan.

SCHEMA OF LUNG INNERVATION

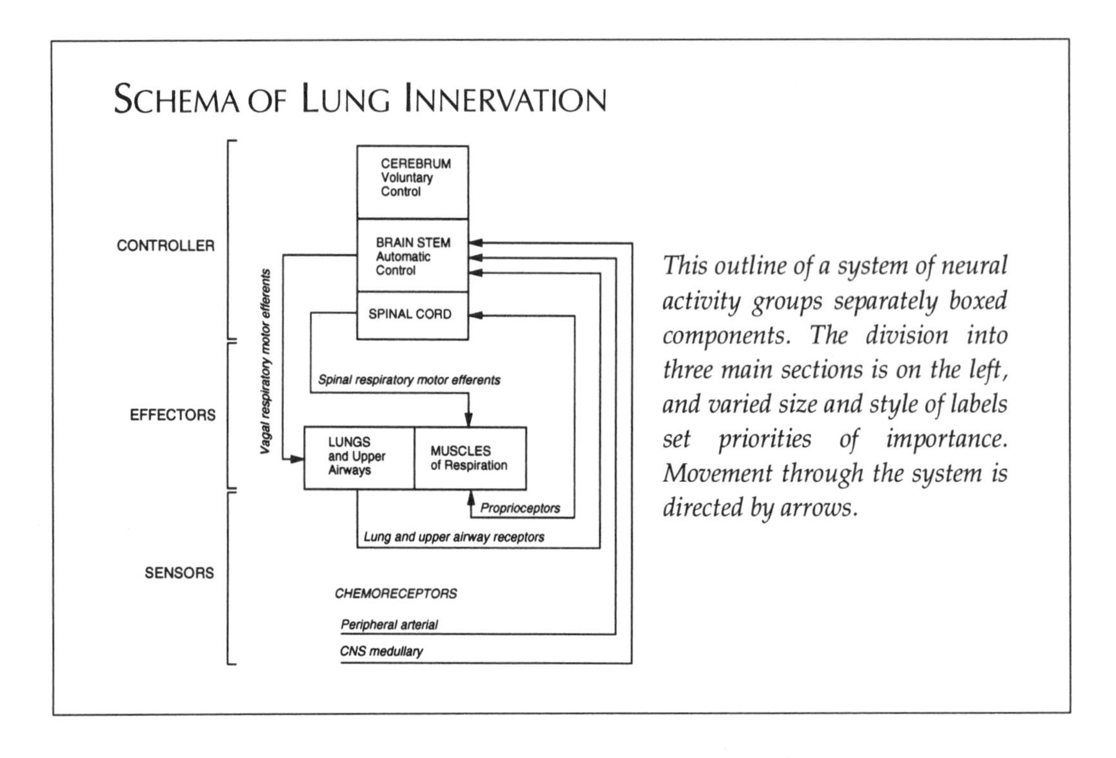

This outline of a system of neural activity groups separately boxed components. The division into three main sections is on the left, and varied size and style of labels set priorities of importance. Movement through the system is directed by arrows.

To communicate, a chart must be carefully organized. Make several rough sketches with different arrangements, keeping in mind the purpose and priorities. Or, if done on a computer, all components can be moved around, eliminated, repeated, enlarged, or reduced. There is software for charting and simple drawing programs for this purpose.

The schema above is much too complex for a slide. This amount of information must be cut or divided into three or more slides. Presentation software such as PowerPoint® may help to simplify information. It has built-in formats for schema, charts, and word slides that impose some restriction on the amount of information.

Whatever method is used, keep the chart or schema simple: do not put too much information in it. Pool ideas with colleagues, and try to make the chart interesting as well as informative. Make use of boxes or circles, simple drawings, arrows, and varieties of size and style of labels.

To be effective, both charts and tables must be clearly and simply organized.

TABLES

Tables, like dictionaries, are indispensable in our lives. What would we do without the table of measurements, the table of contents, the tide table, the train schedule? We all refer to them from time to time for specific information. However, we do not regard them as entertaining, and we do not read through a whole table unless we have to.

Tables for scientific research are often necessary for looking up specific facts. Although they are not interesting reading, good, pleasing design with symmetry, balance, and variety invite the reader to look. Good design can help the reader to find information quickly. Disorganized, poorly designed tables are irritating and visually distressing.

A successful table organizes alphabetical or numerical information into columns and rows that are accurate, logical, and easy to follow and understand. Headings should be plain English and unabbreviated. We should be able to read the table without a magnifying glass.

Use a Table

- To summarize research findings.
- To group together specific data sets to compare and relate.
- To document an account of experimental procedure and results.
- To enable the reader to make calculations from experimental data.
- To enable the reader to reproduce the experiment.

The last two uses for tables, calculation and reproduction of data, although the most important in the larger objective of science, are only useful in published tables. Tables for slides and posters do not allow time for calculation or reproduction uses. Also, because of the short viewing time, tables for posters and slides should contain much less information than those for publication.

Criteria for a Good Table

- A table must be complete; i.e., it must be understandable without detailed reference to the text.
- Use units that are complete and easily understood.
- The data in a table must be not only accurate, but pertinent and significant. Select the important aspects of the information.

- Keep the format clear and simple. Do not waste space, and be consistent in style and terminology.
- Make groupings logical.
- Design the table to fit the format of a slide, page, or poster.

Consider how the viewer will use the information in the table. If the table might be used for calculation or duplication to extend the reader's own work, be sure all data for that purpose are included. For example, an attack on the work of two researchers cited, among other things, the inappropriate choice of statistical analysis and the unavailability of raw data [Research News. Conflict over DNA clock results. Science 241:1598–1600 (1988)]. Present pertinent raw data rather than the statistics derived from them.

However, if the table is being used to make comparisons or as a summary, be more selective about the information.

The table below is complete enough for calculation or possible duplication. It is taken from a journal, and it contains raw data from experiments with five lambs without kidneys (anephric lambs). Calculations have been made to derive the *P*

JOURNAL TABLE

TABLE V

Pulmonary Hemodynamics and Fluid and Protein Transport in Five Anephric, Anesthetized 1–2-Wk-Old Lambs before and after Intravenous Furosemide, 8 mg/kg

	Weight			Vascular pressures		Protein concentrations					
						Lymph		Plasma			
Lamb	Body	Dry bloodless lung	Time relative to furosemide infusion	$\bar{P}pa$	$\bar{P}la$	Total	% Alb*	Total	% Alb*	Lymph flow	Protein flow
	kg	*g*		*torr*		*g/dl*		*g/dl*		*ml/h·g*‡	*mg/h·g·g$_p$*§
15	7.3	19.7	Before	17	7	2.90	55	3.90	49	0.11	0.79
			After	17	7	2.83	55	4.04	47	0.09	0.63
16	6.9	18.9	Before	17	8	2.70	48	4.01	45	0.10	0.66
			After	17	9	2.57	53	3.96	44	0.08	0.51
17	5.4	18.0	Before	21	7	2.40	50	4.03	47	0.10	0.62
			After	21	7	2.60	46	4.19	43	0.08	0.49
18	10.0	24.6	Before	25	−1	2.14	55	3.78	47	0.12	0.65
			After	27	1	2.26	54	3.92	48	0.11	0.62
19	5.2	15.5	Before	21	3	1.95	64	3.78	52	0.17	0.88
			After	21	3	1.91	65	3.76	52	0.15	0.77
Mean±SEM	7.0±0.9	19.3±1.5	Before	20±1	5±2	2.42±0.17	54±3	3.90±0.05	48±1	0.12±0.01	0.72±0.05
			After	21±2	5±1	2.43±0.16	55±3	3.97±0.07	47±2	0.10±0.01	0.60±0.05
P				NS	NS	NS	NS	NS	NS	< 0.05	<0.05

* Percent of total protein concentration which is albumin.
‡ Lymph flow relative to dry bloodless lung tissue.
§ Protein flow (lymph flow × lymph protein concentration) relative to dry bloodless lung tissue and plasma protein concentration.

Horizontal lines divide the title, headings, and footnotes from the body of the table. Complex subheadings are also separated by horizontal lines. The means and P *values are shown below the data. The journal typeset the table to fit exactly into the width of a page, and the style and size of the text were also specified by the journal.*

values; units are complete and described in the footnotes. It contains a large amount of information and is complete enough for calculation or duplication.

The table format shown on the previous page is standard for most journals. Occasionally, some horizontal lines may be thicker or doubled, but vertical division lines are rarely used. Uniform label size is used throughout the table and is usually the same size as the text except when the table is so large it must be reduced. In this table, lettering size is small enough to cause squinting. In addition, the complexity of heading and units of measurement make this a difficult table in which to find specific information.

In the table below, the lamb numbers and weights have been omitted for simplification. Yet it is still complete enough for calculation and reproduction to be made from it. This table also conforms to the journal format.

SIMPLIFIED TABLE FOR A JOURNAL

Data Related to Lung Fluid Balance in Five Anephric, Anesthetized 1-2-wk old Lambs Before and After Furosemide

	VASCULAR PRESSURES		PROTEIN CONCENTRATIONS				FLOW	
			Lymph		Plasma			
Intravenous Furosemide 8 mg / kg	Ppa** Torr	Pla***	Total g / dl	% Alb	Total g / dl	% Alb	% Lymph ml / h g†	Lymph Protein mg / h g gp‡
Before	17	7	2.9	55	3.90	49	0.11	0.79
After	17	7	2.83	55	4.04	47	0.09	0.63
Before	17	8	2.70	48	4.01	45	0.10	0.66
After	17	9	2.57	53	3.96	44	0.08	0.51
Before	21	7	2.40	50	4.03	47	0.10	0.62
After	21	7	2.60	46	4.19	43	0.08	0.49
Before	25	-1	2.14	55	3.78	47	0.12	0.65
After	27	1	2.26	54	3.92	48	01..	0.62
Before	21	3	1.95	64	3.78	52	0.17	0.88
After	21	3	1.91	65	3.76	52	0.15	0.77
Mean±SEM								
Before	20±1	5±2	2.42±0.17	54±3	3.90±0.05	48±1	0.12±0.05	0.72±0.05
After	21±2	5±1	2.43±0.16	55±3	3.97±0.07	47±2	0.10±0.01*	0.60±0.05*

* P<0.05 **Ppa=mean pulmonary arterial pressure ***Pla=mean left atrial pressure
† Relative to dry bloodless lung tissue
‡ Lymph flow x lymph protein concentration / plasma protein concentration, relative to dry bloodless lung tissue

Information for tables should be simplified as much as possible. Leave out data that have no bearing on the point you want to make. Here, the range or means of the lamb weights could be stated in one sentence. The P values are indicated by asterisks. The values without asterisks, by implication, are not significant. Some information in the title has been moved to the heading and the units are put directly under the headings that they describe. Lettering is larger and easier to read.

The table is submitted to a journal in typewritten form and the journal resets it in its own format for publication. If you think it helps your information to use a different format, you could and should request this. Below is the same table in a format that does not conform to the journal's specifications but that has clearer divisions, making it easier to locate data.

CLEARLY ORGANIZED TABLE

Data Related to Lung Fluid Balance in Five Anephric, Anesthetized 1-2-wk old Lambs Before and After Furosemide

Intravenous Furosemide 8 mg / kg	VASCULAR PRESSURES		PROTEIN CONCENTRATIONS				FLOW	
	Torr		Lymph		Plasma			
	Ppa	Pla	Total g / dl	Alb	Total g / dl	Alb	% Lymph ml / h g†	Lymph Protein mg / h gp‡
Before	17	7	2.90	55	3.90	49	0.11	0.79
After	17	7	2.83	55	4.04	47	0.09	0.63
Before	17	8	2.70	48	4.01	45	0.10	0.66
After	17	9	2.57	53	3.96	44	0.08	0.51
Before	21	7	2.40	50	4.03	47	0.10	0.62
After	21	7	2.60	46	4.19	43	0.08	0.49
Before	25	-1	2.14	55	3.78	47	0.12	0.65
After	27	1	2.26	54	3.92	48	0.11	0.62
Before	21	3	1.95	64	3.78	52	0.17	0.88
After	21	3	1.91	65	3.76	52	0.15	0.77
Mean ± SEM								
Before	20 ± 1	5 ± 2	2.42±0.17	54 ± 3	3.90±0.05	48 ± 1	0.12 ± 0.01	0.72 ± 0.05
After	21 ± 2	5 ± 1	2.43±0.16	55 ± 3	3.97±0.07	47 ± 2	0.10 ± 0.01*	0.60 ± 0.05*

* P <0.05
† Relative to dry bloodless lung tissue
‡ Lymph flow x lymph protein concentration / plasma protein concentration, relative to dry bloodless lung tissue

The use of vertical lines clarifies the divisions, and the use of upper case and bold lettering sets unambiguous priorities of headings and subheadings. In a table as complicated as this one, use every means possible to clarify the information.

Unfortunately, a request to use a different format may not be granted by the journal. If such is the case, more information might be eliminated so that, even in conformance with the journal standards, the table will be easy to understand.

When simplifying by eliminating information, consider carefully the purpose of the table. If the intention is to summarize findings, the use of means and standard errors would be most effective. If the findings are to be compared and related, use only the pertinent data sets. Be selective about the number of data sets if documentation or facilitation of calculation is the goal. For reproduction of the experiment, two or more tables may be better than one if the information is long and complicated.

TABLES FOR SLIDES

The conclusions to be drawn from the above tables, as stated in the text of the paper, are that although pressures and concentrations do not change, lymph flow decreases by 17% in lambs without kidneys (anephric lambs).

For a slide, these conclusions must be emphasized, and the table must be further simplified. It is easier to see that protein concentrations are the same without the albumin data. Also, unless each of the five lamb studies will be discussed separately, it is more effective to show the mean values only. It is here that the significance of the decrease in flow is most evident.

For best results, the table should be completely redesigned for use as a slide (below). The means and standard errors only are used. The change in lymph flow is emphasized by eliminating lymph protein flow and by making lymph flow the first

TABLE FOR A SLIDE

LYMPH FLOW DECREASES IN ANEPHRIC LAMBS AFTER FUROSEMIDE

Intravenous Furosemide 8 mg / kg	**LYMPH FLOW** ml / h g	**VASCULAR PRESSURES** Ppa (Torr)	Pla (Torr)	**PROTEIN CONCENTRATIONS** **Lymph** (g / dl)	**Plasma** (g / dl)
BEFORE	0.12 ± 0.01	20 ± 1	5 ± 2	2.42 ± 0.17	3.90 ± 0.05
AFTER	0.10* ± 0.01	21 ± 2	5 ± 1	2.43 ± 0.16	3.97 ± 0.07

*P < 0.05

By cutting out all but the essential information, the text in the slide is larger and easier to read. The title is changed and abbreviated to state the most important conclusion. The standard errors have been put below the means to conform better to the rectangular format. The speaker could explain the lymph protein units if necessary. Avoid footnotes in a slide.

column. Because we read from left to right, the first column on the left is the most prominent.

Keep in mind that a slide is on the screen for a short time. Use this limited time to communicate the most important facts.

> For a slide of a table, use only concise and pertinent information.

It is not unusual to see a lecture slide of a table, probably taken from a journal, that is so full of minuscule text that it might as well be hieroglyphics. The lecturer then proceeds to tell us to disregard most of the table and to concentrate on one part (which we cannot see). The result is boredom and/or irritation. If a table must be used in a talk, take the time to design one that can be easily seen and understood.

However, for a talk, it might be more concise and informative to design a graph to show the significant observations. (The fact that there was no change in pressure and concentration may be said in one sentence.) Two observations could be

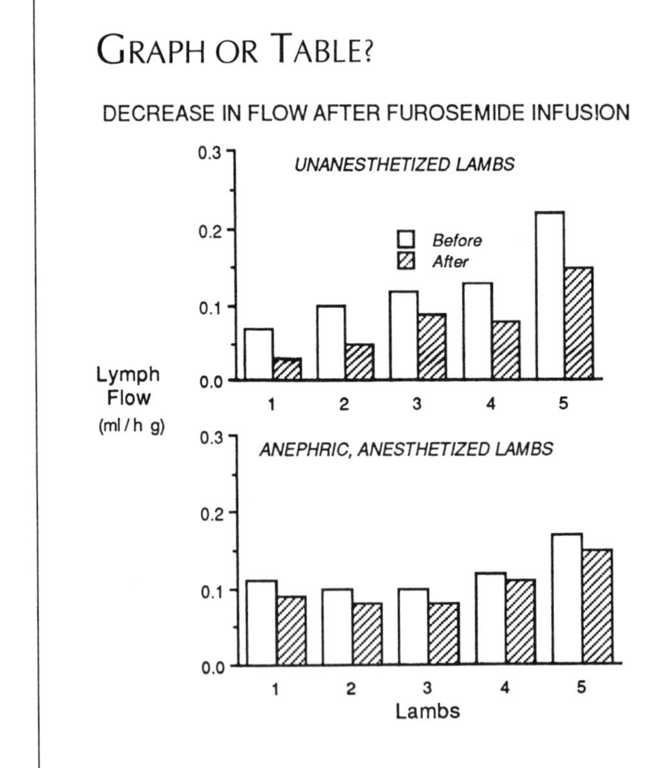

In this graph, two points from the previous table are illustrated: (1) Lymph flow decrease after furosemide is shown by the before and after bars. (2) Greater decrease of flow in unanesthetized lambs is clear in comparing the top and bottom graphs. Five experiments were done with both groups, and it can be seen that there was less lymph flow in general among the anephric lambs.

illustrated in one graph. First, lymph flow decreased in lambs with no kidneys: second, lymph flow decrease was greater in unanesthetized lambs with kidneys.

Tables, especially complex tables, are not effective slides. Think carefully about how you want to present the information, and before making final decisions, experiment with different forms and different ways to organize. Above all, be clear about what you want to communicate in your table.

Consider the possibility of using a drawing with a table. If the data relate to anatomical areas as in the following table, values will be quickly associated and remembered more easily.

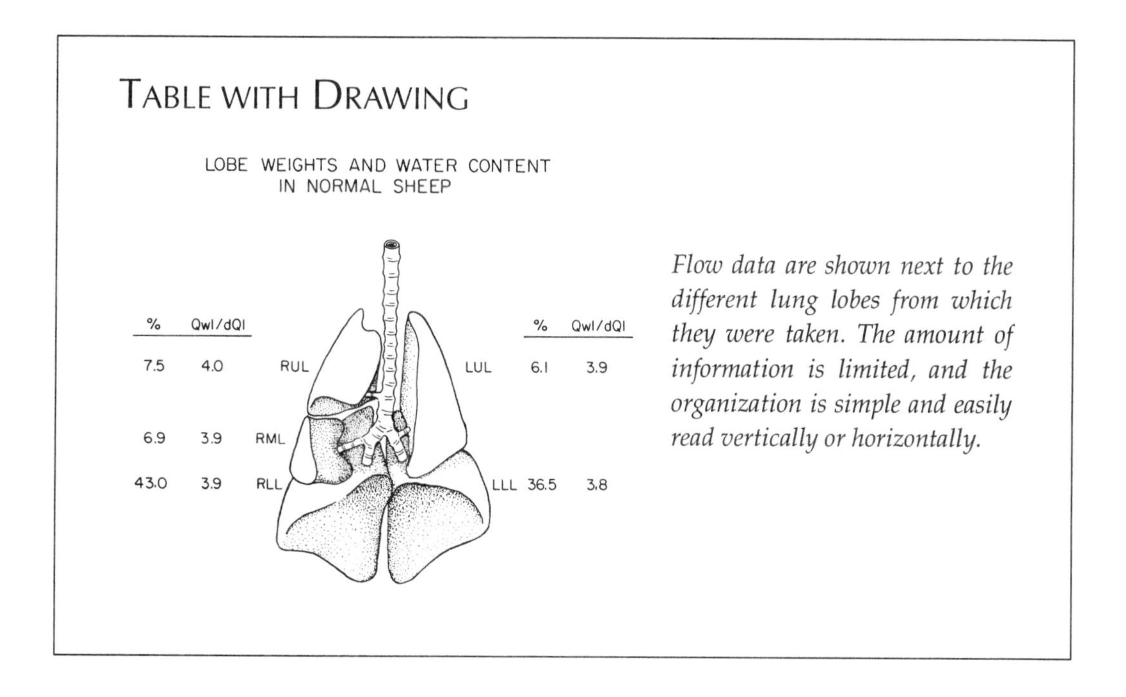

Flow data are shown next to the different lung lobes from which they were taken. The amount of information is limited, and the organization is simple and easily read vertically or horizontally.

The greatest pitfall in designing tables is to include too much information. A page or screen full of lists is daunting and discouraging to the viewer and is not helpful in communicating. Every table can benefit from discriminating selection and cutting.

To Condense Large Tables:

- Remove unnecessary rows and columns. Select the important columns and leave out the rest.
- Combine columns that are closely related by using slashes or dashes. For instance, separate columns such as, Total Protein and % Albumin in protein

could become Total / % Albumin. The data under that heading would then be written 2.90 / 55.
- Do not repeat data that are already in the text or in another table. Do not repeat label information in the heading that may already be in the title or footnote.

Before making a table, ask yourself if it is absolutely necessary. Could that information be said in a few sentences in the text? Could that information be made into a graph? If a table is necessary, limit the amount of information in it and organize it carefully.

5 MOLECULAR GRAPHICS

The biological story told in molecular structure is still new, exciting, and full of mystery and challenge. Graphical representation of sequences of nucleotides in DNA and RNA, and of amino acids in proteins, and the stereoscopic display of their structure, has played a crucial part in understanding as well as communicating this subject. The graphics used in biochemical study are sequences, sequence alignment and folding, maps of nucleic acid sequence restriction sites, and models of molecular structure.

In this chapter, the focus is on black and white representation of the graphics because that is still most widely used in journals and slides. Color, however, is so helpful in clarifying the complexities of models, in particular, that a discussion of its use is briefly mentioned.

GENETIC SEQUENCES

A sequence is a list of the residues in the primary structure of a molecule. As new sequences are determined, these data may be fed into a computer. Computer programs have been developed that include protein data banks. These programs translate DNA and RNA sequences into amino acids. They find homology and consensus in aligning different sequences, and they locate patterns or motifs. Possibilities for the folding of the polypeptide chains may be displayed, and sites for enzyme cutting or restriction may be mapped.

Sophisticated computer programs have been developed for analyzing and displaying molecular sequences. Not only do such programs digest and analyze great quantities of information, they do it rapidly and accurately. These programs are tools for the researcher, and the graphical output from them is often used in publications, slides, and journals.

Entire sequences are now rarely shown in journals unless they are new or offer new insights. Growing data banks of sequences are now available for those who want to look them up. But often regions of a sequence are shown and generally areas of interest are shown by graphical devices such as underlining, boxing, and arrows and texture.

Raw computer-printed figures from such programs may not be effective for communicating information because they may be too small to be read, too complex to be understood, and the wrong shape to fit the medium used. The following examples suggest ways to modify computer-printed sequences to communicate more effectively. Some sequencing software has options for making these changes; others allow transfer to a drawing program to make changes and additions. Changes may also be made by hand.

The sequence shown on the following page is a working blueprint. From it can be discovered the number and length of repeated motifs and the sites at which DNA excision enzymes may act. It is a tool for working and deduction. In its raw form, however, it discourages communication. The explanation for such a figure would take a great deal of space and time.

If it is necessary to show a whole sequence, treat it as an orientation map and emphasize areas of interest. If necessary, extrapolate areas of interest by making a separate figure. This may entail enlargement and addition of lines, boxes, and patterns. To achieve these graphical options, it may be necessary to transfer the sequence to a computer drawing program. Or it may be necessary to cut and paste and add graphical options and labels by hand. (See Chapter 12, "Drawing by Hand.") Although this takes time and thought, if it gets the message across, it will be worth it.

Sequences may be clarified and simplified by changes in spacing and use of graphical devices. Spacing is an important option to consider. On the following pages are examples showing how numbering and spacing may enhance readability.

A sequence requires a font that is monospaced (each letter takes up the same amount of space) since the alignment of one letter above the other must be exact. Courier or Monaco or GeneFont (LaserGene software by DNASTAR, Inc.) are

TRANSLATED SEQUENCE

```
-35                 -30
Asn Thr Thr Thr Gly Glu Ser Ala Asp Pro Val Thr Thr Thr Val
AAC ACT ACC ACC GGG GAG TCT GCA GAC CCT GTC ACC ACC ACC GTT

-20                                     -10
Glu Asn Tyr Gly Gly Glu Thr Gln Val Gln Arg Arg Gln His Thr
GAA AAC TAC GGC GGT GAG ACA CAA GTC CAA CGA CGT CAG CAC ACC

                      1                                  10
Asp Val Thr Phe Ile Met Asp Arg Phe Val Lys Ile Gln Asn Leu
GAC GTT ACT TTC ATA ATG GAC AGA TTT GTA AAG ATA CAA AAT TTG

                                     20
Asn Pro Ile His Val Ile Asp Leu Met Gln Thr His Gln His Gly
AAC CCC ATA CAT GTC ATT GAC CTC ATG CAA ACC CAC CAA CAC GGG

                 30                                      40
Leu Val Gly Ala Leu Leu Arg Ala Ala Thr Tyr Tyr Phe Ser Asp
TTG GTA GGT GCC CTG TTA CGT GCT GCT ACG TAC TAC TTC TCT GAC

                                     50
Leu Glu Ile Leu Val Arg His Asp Gly Asn Leu Thr Trp Val Pro
CTG GAG ATT CTG GTA CGC CAT GAC GGT AAC CTA ACC TGG GTA CCC

                 60                                      70
Asn Gly Ala Pro Glu Ala Ala Leu Ser Asn Met Gly Asn Pro Thr
AAC GGA GCA CCC GAG GCA GCT CTG TCT AAC ATG GGC AAC CCC ACC

                                     80
Ala Tyr Pro Lys Ala Pro Phe Thr Arg Leu Ala Leu Pro Tyr Thr
GCC TAC CCC AAG GCA CCA TTT ACG AGG CTC GCG CTC CCC TAC ACC

                 90                                     100
Ala Pro His Arg Val Leu Ala Thr Val Tyr Asn Gly Thr Ser Lys
GCG CCA CAC CGC GTA TTG GCG ACA GTG TAC AAC GGG ACG AGC AAG

                                    110
Tyr Ser Ala Gly Gly Met Gly Arg Arg Gly Asp Leu Glu Pro Leu
TAC TCC GCA GGT GGT ATG GGC AGA CGG GGC GAC CTA GAG CCT CTC

                120                                     130
Ala Ala Arg Val Ala Ala Gln Leu Pro Thr Ser Phe Asn Phe Gly
GCG GCG AGG GTC GCC GCT CAG CTT CCT ACT TCT TTC AAC TTT GGT

                                    140
Ala Ile Gln Ala Thr Thr Ile His Glu Leu Leu Val Arg Met Lys
GCA ATT CAA GCC ACG ACC ATC CAC GAG CTC CTC GTG CGC ATG AAG
```

On the left is part of a DNA alpha 22 sequence with its translation from nucleotide base to amino acid residue. Grouping indicates three-letter codons. Above the sequence, every tenth codon is numbered. This sequence has been reduced to the size of one journal column.

examples of monospaced fonts. Fonts such as Times or Helvetica are proportionately spaced, so that some letters take up more space than others. Upper and lower case letters of monospaced fonts take up the same amount of space. The nucleotide bases in the previous figure use upper and lower case but translation to the all-upper-case amino acids is still exactly in line. Software such as GeneWorks® from IntelliGenetics, Inc., allows the choice of fonts, size, and style.

Compare the two figures on the following page.

Numbering of sequences may be done above the sequence line or at the beginning or end of the lines (both shown in the upper figure). Spacing between lines of the sequence should be great enough to indicate distinct separation but not so much as to cause discontinuity. If arrows and asterisks or other graphic devices are used, it may be necessary to leave more space between the lines. Many programs allow for such space adjustments.

VERTICAL SPACING

```
        10        20
12345 67890 12345 67890 12345
-------------------------------------
CTTGT CCTCC TGGTA TTGGG ATTTG      25
AGGTC CAGGG GACCC AACAG CCCCA      50
GCAAG ATGAG ATGCC TAGCC CGACC      75
TTCCT CACCC AGGTG AAGGA ATCTC     100
TCTCC AGTTA CTGGG AGTCA GCAAA     125
GACAG CCGCC CAGAA CCTGT ACGAG     150
AAGAC ATACC TGCCC GCTGT AGATG     175
AGAAA CTCAG GGACT TGTAC AGCAA     200
AAGCA CAGCA                       210
```

The figure on the left is narrow because there are 25 bases in each line. This fits well into a two-column journal. For a one-column journal or for a slide, there would be wasted space to the right and left of the figure.

Both sequences on this page are ApoC2 cDNA and were generated in the GeneWorks® program.

HORIZONTAL SPACING

```
        10         20         30         40         50
1234567890 1234567890 1234567890 1234567890 1234567890
----------------------------------------------------------
CTTGTCCTCC TGGTATTGGG ATTTGAGGTC CAGGGGACCC AACAGCCCCA    50
GCAAGATGAG ATGCCTAGCC CGACCTTCCT CACCCAGGTG AAGGAATCTC   100
TCTCCAGTTA CTGGGAGTCA GCAAAGACAG CCGCCCAGAA CCTGTACGAG   150
AAGACATACC TGCCCGCTGT AGATGAGAAA CTCAGGGACT TGTACAGCAA   200
AAGCACAGCA                                               210
```

Fitting 50 paired bases in each line results in a wide figure. The wide or more rectangular shape is appropriate for a one-column journal in which one column fills the page or for a slide that fits the screen best in a wide format.

Analysis of the sequence is often the purpose of the figure. Graphical ways of showing analysis are indicated below.

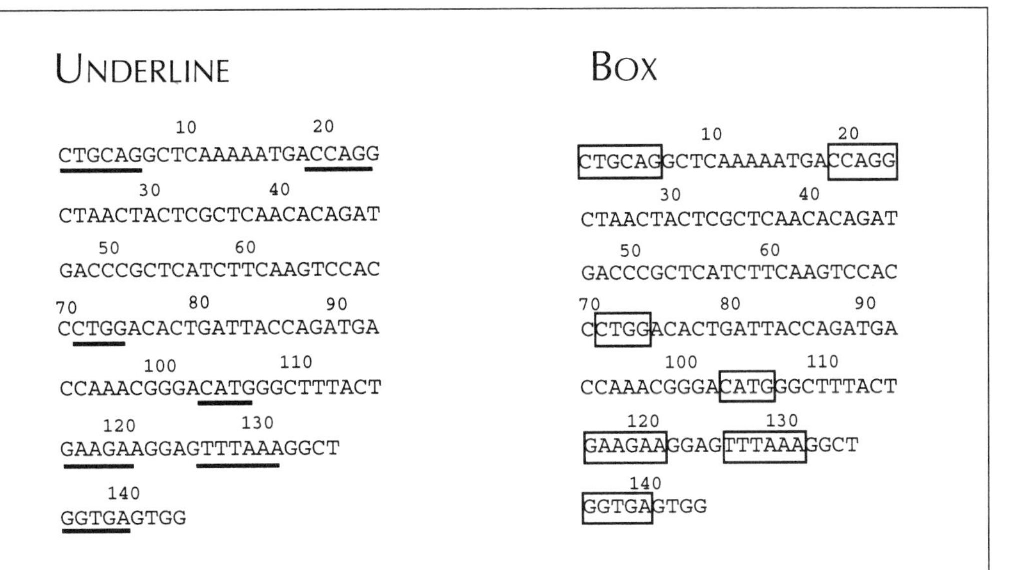

Repeats of nucleotides, regions of sequences, and sites of enzyme cleavage are some aspects of sequence analysis. Above, enzyme cleavage sites are indicated in different ways. Underlining is the simplest and clearest. Boxing groups of letters is effective, but because the letters are so close together, the boxes may distort or obscure the letters themselves.

Sometimes it is necessary to make several distinctions in analysis. In this case, there are other graphical possibilities. (See figures on page 54.)

If the program for generating sequences does not have any graphical options, consider exporting the data of the sequence to a drawing program to make changes and additions. Another choice, simpler for some, is to make the additions by hand. (See Chapter 12, "Drawing by Hand.")

Whole sequences are rarely shown in talks or publications because most known sequences are accessible from data banks. New sequences are shown, and sequence comparisons or alignments are used to show similarities or differences. Alignment is shown on page 55 by using dashes to indicate gaps, upper case letters to indicate exact matches, and lower case letters to indicate no match.

OVERLINING, UNDERLINING, AND LINE STYLES

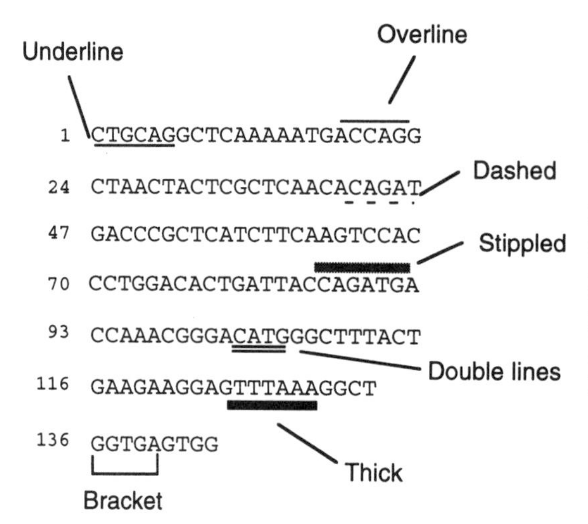

Selection of areas of a sequence is shown in various ways here. Overlining is distinguishable from underlining. Dashed lines are the least conspicuous. Stippled bars and double lines tend to fuse when reduced. Thin or thick lines are easy to distinguish, and brackets show groupings clearly.

SHADING, ARROWS, AND ASTERISKS

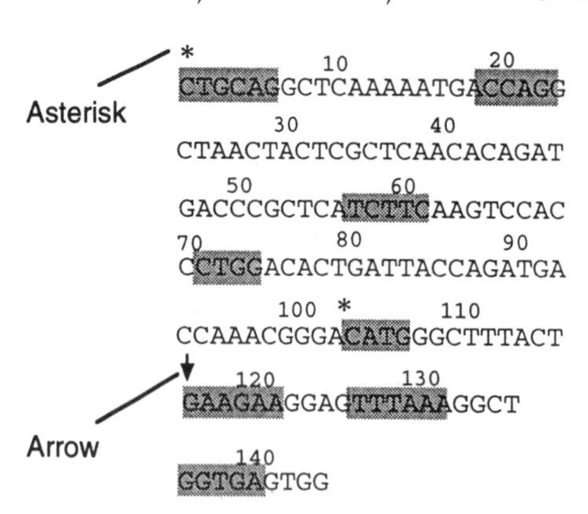

For shading, use a light stipple pattern (20% or less) because shading can obscure the letters to be emphasized. In reduction, the dots of the shading tend to fuse together and become solid black. Make arrows and asterisks large enough to be seen and easily distinguished in reduction. The MacDraw Pro® program was used to make the changes in the figures on this page and pages 53 and 55.

SEQUENCE ALIGNMENT

```
1         GDvekGkKIFimKCsqCHTVekGgkHKtGPNLhGLFGRktGqapGYSYtAANKNKgiiWgedTL
          ||   | |||  ||  ||||  |  || |||| |||||  |   |||| ||||||   |   ||
1 asfaeappGDkdvGgKIFktKCazCHTVzlGagHKqGPNLnGLFGRqsGttaGYSYsAANKNKavlWabbTL

  --------GD---G-KIF--KC--CHTV--G--HK-GPNL-GLFGR--G---GYSY-AANKNK---W---TL


65 meYLeNPKKYIPGTKMiFvGiKKkeeRADLIAYLKkATne
     || ||||||||||| | | ||   ||||||||| ||
73 ydYLlNPKKYIPGTKMvFpGlKKpqdRADLIAYLKhATa

   --YL-NPKKYIPGTKM-F-G-KK---RADLIAYLK-AT--
```

Shown above is a sequence alignment of human and parsnip cytochrome amino acids. Here amino acids are represented by single letters rather than a three-letter abbreviation. Although this is more compact, quick identification is hampered. The alignment is shown as it appears on the computer screen. Capital letters joined by vertical dashes denote exact alignment. The horizontal dashes in the bottom line denote gaps in alignment.

To make this figure readable and more informative, it should be rearranged and fully labeled.

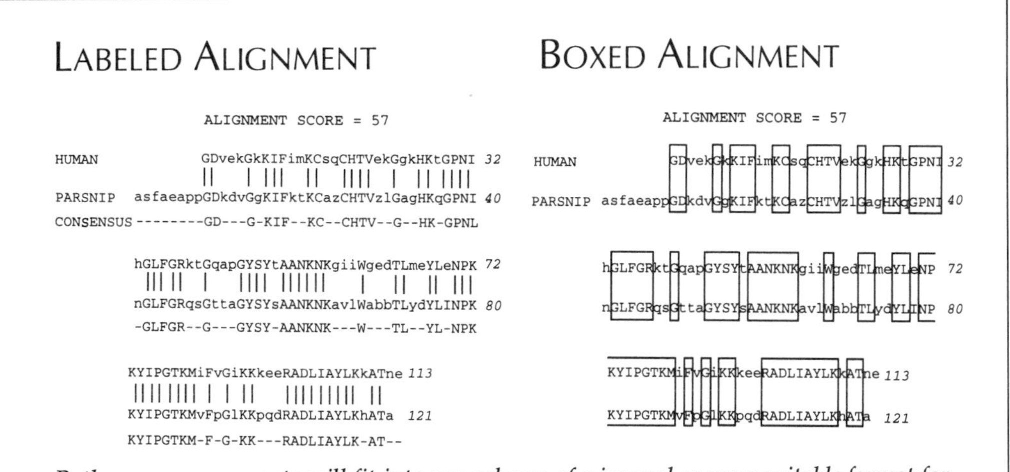

Both new arrangements will fit into one column of a journal or are a suitable format for a slide. In the figure on the left, the three rows are labeled and the consensus score is given in the title. On the right, alignment is indicated by boxing. This eliminates the need for a consensus line and also makes the areas of homology visually clearer. These changes can be made by hand or in a drawing program.

Programs differ in their manner of displaying nucleotide sequence alignment. Below are alignments of ApoC2 cDNA. The alignment on the left is from GeneWorks® and shows enzymes that cut no more than two times. The alignment on the right is from DNA Strider™ and shows endonuclease sites.

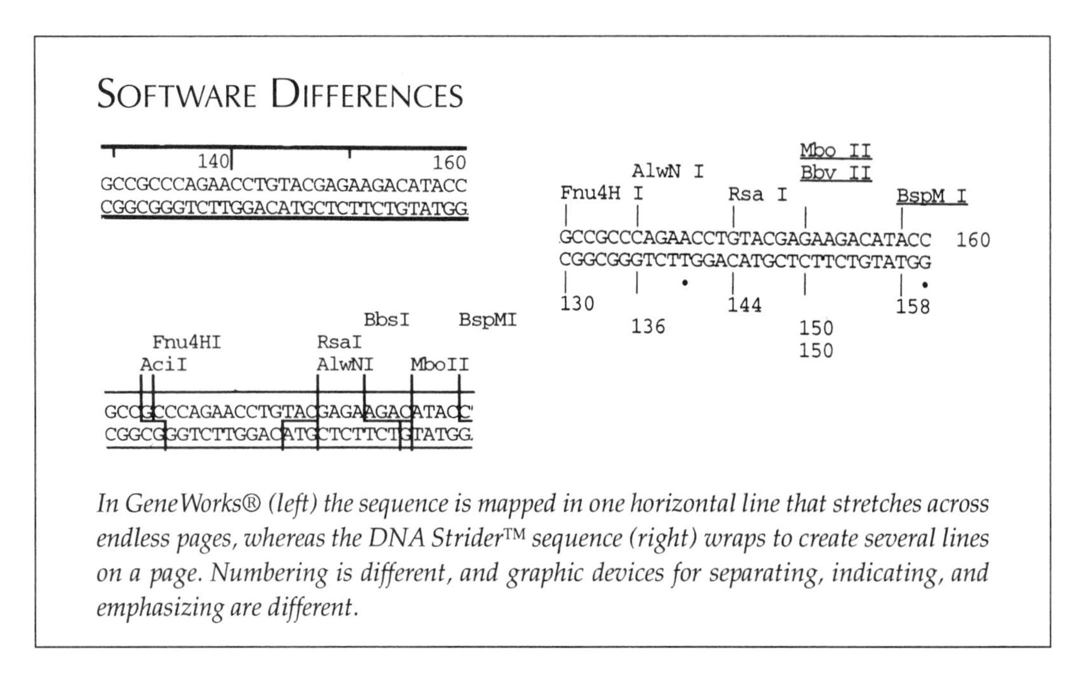

In GeneWorks® (left) the sequence is mapped in one horizontal line that stretches across endless pages, whereas the DNA Strider™ sequence (right) wraps to create several lines on a page. Numbering is different, and graphic devices for separating, indicating, and emphasizing are different.

Below are some guidelines that will make sequences more intelligible and effective for communicating information.

Be clear about the purpose when preparing sequences for publication, slides, or posters.

- To introduce a new sequence and translation, the entire sequence must be shown.
- For comparison of sequences, it might be more effective to show *only* the regions of greatest interest.
- To support other findings, use *only* what is pertinent to the finding.

Design the sequence for the medium.

- A slide of a long sequence will be meaningless to most viewers. Select only parts of a sequence for a slide. Make the labels complete and large and include a title.
- Short sequences are also better for posters, and for this purpose, it is especially important that labels be large.

- Although more information can be included in a sequence for publication, it will be trying and tedious for the reader to wade through a page of sequences and legends to find the points of interest. When possible, select parts of the sequence that focus on the points you want to make.
- Use an appropriate format. A horizontally aligned sequence is better for a slide than one that is vertically aligned. For publication, a sequence can be planned for long and narrow column width size, or it can stretch across the width of a page, covering a whole page or part of the page.
- Avoid single-letter designations for amino acids unless space requirements force this cryptic state. Three-letter amino acid designations communicate more effectively to a wider audience.

Do you really need to show the sequence?

- Can the point of interest of a sequence be stated in a sentence or two?
- Will a map, model, graph, or gel emphasize the point as well or better than a sequence?
- Remember that the data of the sequence, unless it is very new, is deposited in an accessible database that is available to the viewer.

Use graphical elements carefully.

- Be sure that elements are distinguishable from one another.
- Be sure that elements such as arrows or asterisks are large enough to be seen.
- Be sure that the devices used to clarify do not obscure or distort.

RESTRICTION MAPS

Restriction maps show the sites of restriction enzyme cleavage. They may also show the lengths of the fragments. They are helpful in determining the structure of genes and in describing cloned fragments.

On the next page, the information in the upper sequence is mapped linearly in the lower one.

In the lower restriction map, the sequence line is not continuous; the enzymes actually cut the line. Although this may be symbolically accurate, the sequence line itself is diminished in importance. Also there is a general impression of frequency and relative position of the enzymes, but the locations of the restriction sites are not exact.

SEQUENCE SHOWING RESTRICTION SITES

```
 1    Cys Arg Leu Lys Asn Asp Gln
    C TGC AGG CTC AAA AAT GAC CAG
    |                     |
  PstI                  EcorII
                        ScrGI

 23 Ala Asn Tyr Ser Leu Asn Thr
    GCT AAC TAC TCG CTC AAC ACA

 44 Asp Asp Pro Leu Ile Phe Lys
    GAT GAC CCG CTC ATC TTC AAG
                     |
                   MboII

 65 Ser Thr Leu Asp Thr Asp Tyr
    TCC ACC CTG GAC ACT GAT TAC
                         |
                       EcorII
                       ScrFI

 86 Gln Met Thr Lys Arg Asp Met
    CAG ATG ACC AAA CGG GAC ATG
                         |
                       NlaIII

107 Gly Phe Thr Glu Glu Glu Phe
    GGC TTT ACT GAA GAG GAG TTT
              /    /     /   \
         MboII MnlI AhaIII  MseI

128 Lys Arg Leu Bal Ser
    AAA AGG CTG GTG AGT GG
             |
           HphI
```

The sites of enzymatic cleavage and enzyme names are added to this DNA strand. This works well for a small sequence. For longer sequences and for a more diagrammatic approach, a scaled map works better.

MAP SHOWING RESTRICTION SITES

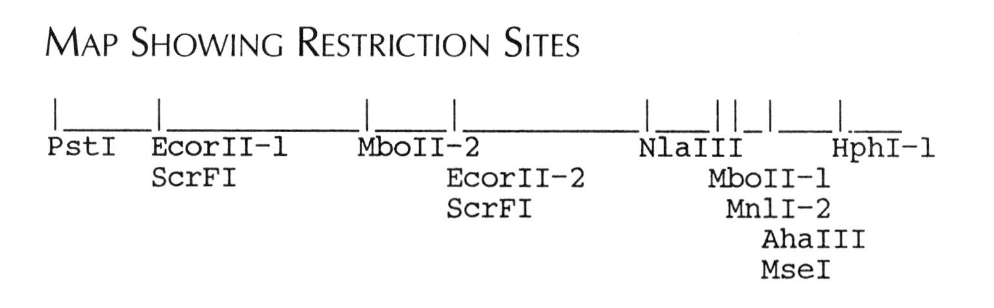

This figure shows the restriction map as it comes from the computer. The sequence is represented by the horizontal line. The enzymes are labeled and the cleavage sites are indicated by the vertical lines.

Some programs allow for continuous lines or thicker lines, larger labels, and different fonts. Or the map may be labeled more clearly and fully by transferring it to a graphic program or drawing it by hand.

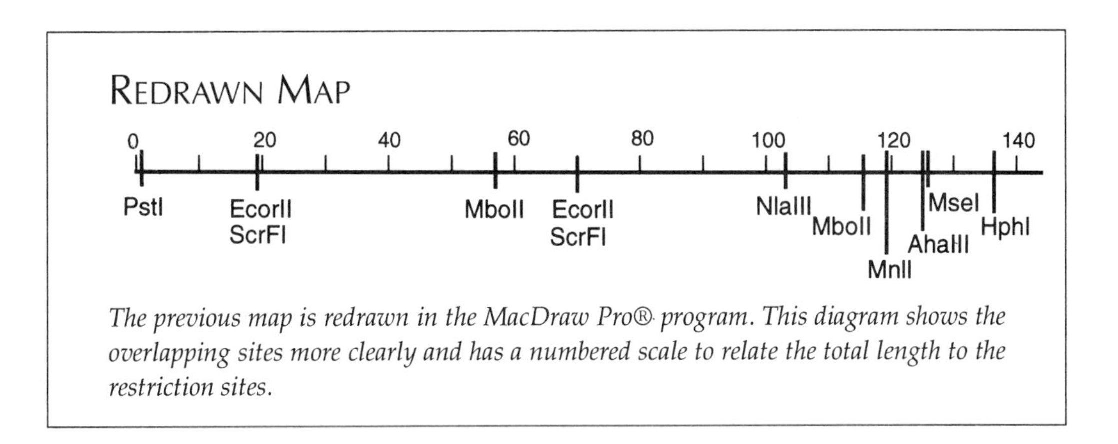

The previous map is redrawn in the MacDraw Pro® program. This diagram shows the overlapping sites more clearly and has a numbered scale to relate the total length to the restriction sites.

Restriction maps are often combined with other elements such as the sequence of a particular restricting enzyme or repeated patterns or motifs shown below.

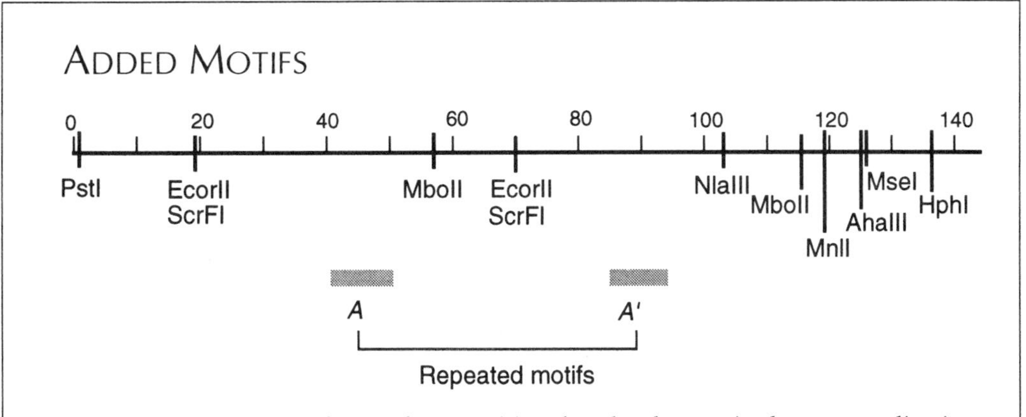

A' is a repeat of the motif A. Both are positioned under the area in the sequence line in which they appear. Because the repeat is of secondary importance in this figure, the motif areas are indicated by stippled bars, which are less obtrusive than solid lines and also symbolize areas. Labels for the motifs are the same size as the enzyme labels, but italicized to make them different.

When displaying different kinds of information in the same figure, make sure the differences are apparent visually. This can be done by spatial separation and the use of contrasting graphical devices used such as textures, line weight, and differences in label size and style.

Circular Maps

Maps may also be shown in circular form. Plasmids (small circular segments of DNA) are best shown this way.

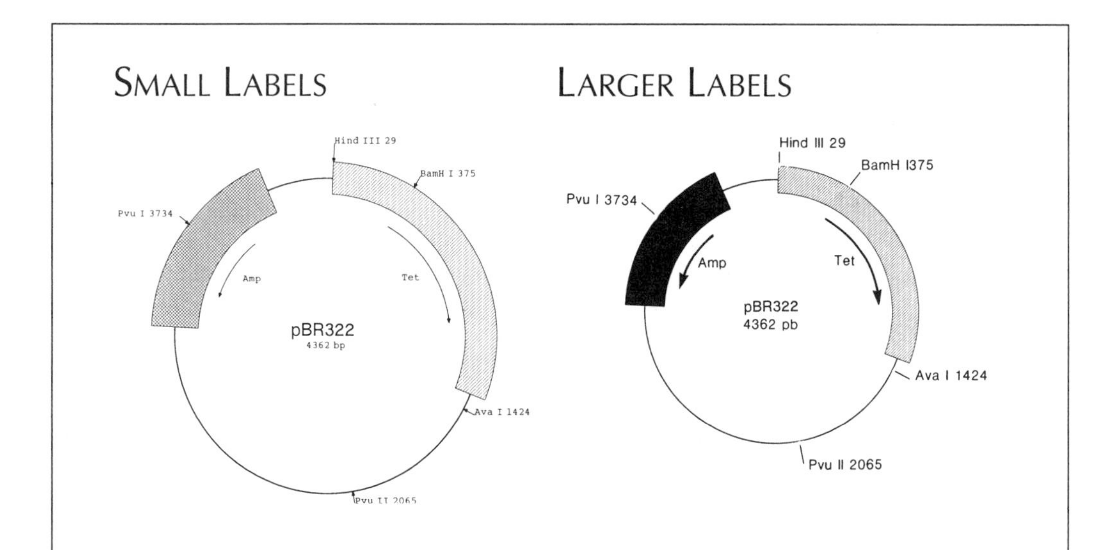

On the left is a plasmid map that shows sites for enzymatic cleavage. Regions of interest are demarcated by the hatched and crosshatched areas. Both of these patterns are too fine lined and similar to be easily seen in reduction. Labels are hard to read. On the right, enzyme labels and arrows were enlarged, and "Amp" texture is solid black to distinguish it from "Tet." New and larger labels, arrows, and darkened texture for this figure are possible in some programs or may be done by hand.

Open Reading Frame Map

An open reading frame (ORF), indicating DNA-to-protein translation and its reverse, is shown by using the symbolism of arrows pointing in different directions. The ORF on the following page shows the ApoC3 gene in the GeneWorks® program.

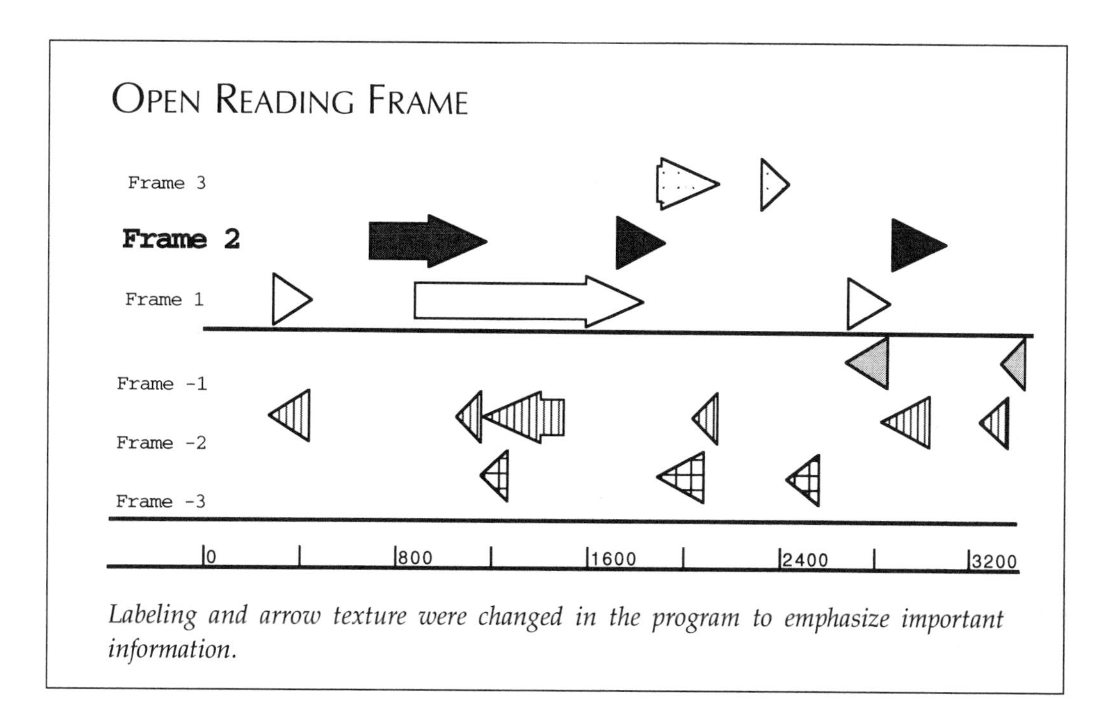

Labeling and arrow texture were changed in the program to emphasize important information.

Helical Net Diagram

Another analytical tool is the helical net. The diagram on the next page shows alpha helical regions of protein sequences.

Analysis Plot

An analysis plot is based on an algorithm table. Most programs have many sequence analysis algorithms. The plot on the next page shows base frequencies or how rich in G/C default bases a sequence is.

Mapping is a compact way to show large amounts of information and is a useful analytical tool. In fact, most maps are intended more for analysis than for display and communication of information. If a map will be used for a publication, slide, or poster, look carefully at label size, line thickness, symbol proportion and texture. The guidelines for effective sequences could also be used for maps. (See page 56.)

- Make the labels large enough to be read easily.
- Design the map as much as possible to conform to the medium used.
- Make use of graphic tools such as thick and thin lines, arrows, brackets, and shading, if necessary, by transferring to a drawing program or by hand.

Helical Net

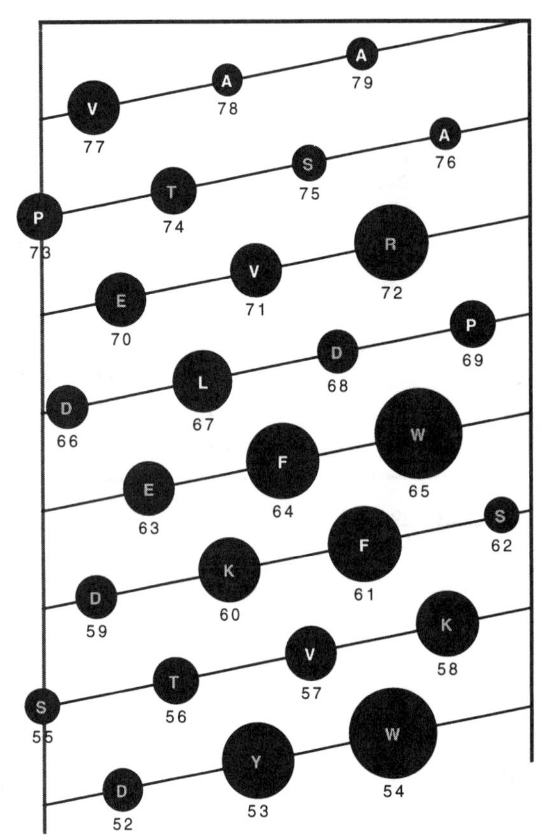

This helical net is part of the ApoC-3cDNA gene. Parameters for residue sorting and angle can be changed.

Analysis Plot

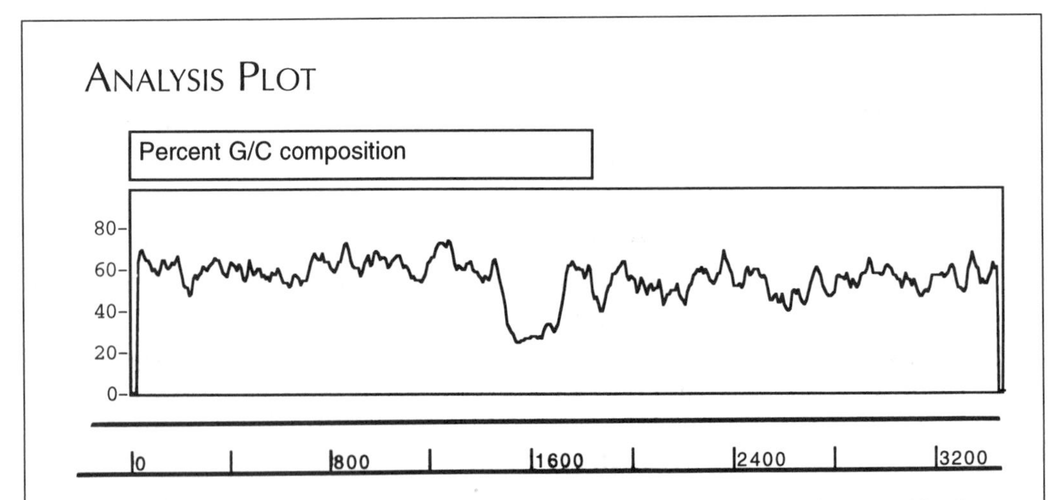

This graph shows a kind of DNA footprinting of the ApoC3 gene. The lows in this plot show A-T-rich bases. Both figures on this page were produced by the GeneWorks® program.

Tables and Graphs

Tables of information are available in sequencing software to show residue number, type of amino acid, hydrophobicity, features, restriction and peptidase sites, reading frames, and more. These tables are meant to be consulted in analyzing sequences. They are not meant for display. If a table is necessary for a figure, use only pertinent information and enter it into a word processing program for graphical changes such as table format, label size, font, and style. (See Chapter 4, "Charts and Tables.")

Scattergrams and histograms, densitometer graphs, and graphs showing hydrophobicity of protein molecules represent useful aspects of the sequence and are possible in most programs. Programs differ in their capabilities for graphical change of plots, so hand work or transfer to a drawing program may be necessary. For example, the isoelectric graph below needs larger, horizontal labels, elimination of the grid, and darkening of the curve.

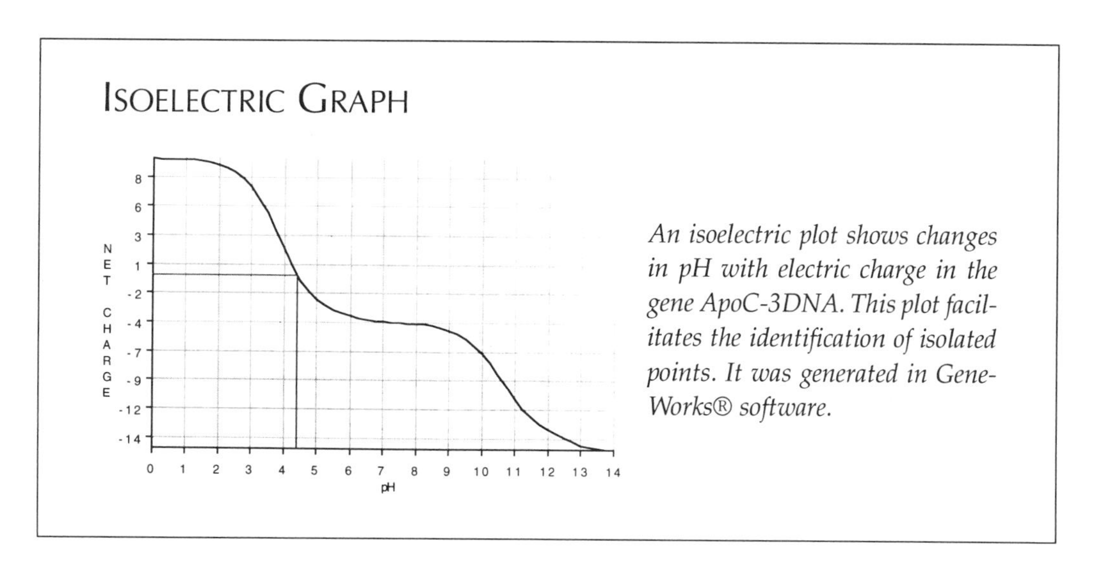

An isoelectric plot shows changes in pH with electric charge in the gene ApoC-3DNA. This plot facilitates the identification of isolated points. It was generated in GeneWorks® software.

MOLECULAR MODELS

Models are two- or three-dimensional diagrams of molecular structure. They are useful for study and analysis of molecules and for visualizing their spatial arrangement and interaction with other molecules. Many complex models are computer generated using data derived from sequencing and mapping. Artists are also extensively involved in model design and are still coming up with new ways to display molecules. Their newest tool is the computer, with its increased choice of modeling software.

Although some models are generated from sequences, other models of molecular structure are discovered through X-ray crystallography. The structure of a molecule may reveal function. It may reveal how it interacts with other molecules. Although the shapes of molecules are hypothetical, they nevertheless allow us to model what is going on at a molecular level. The understanding of molecular structure relies heavily on pictorial presentation.

One of the simplest, and yet the most crucial, models derived from the sequence is the pattern of folding. Folding is the basis for understanding the structure of the molecule.

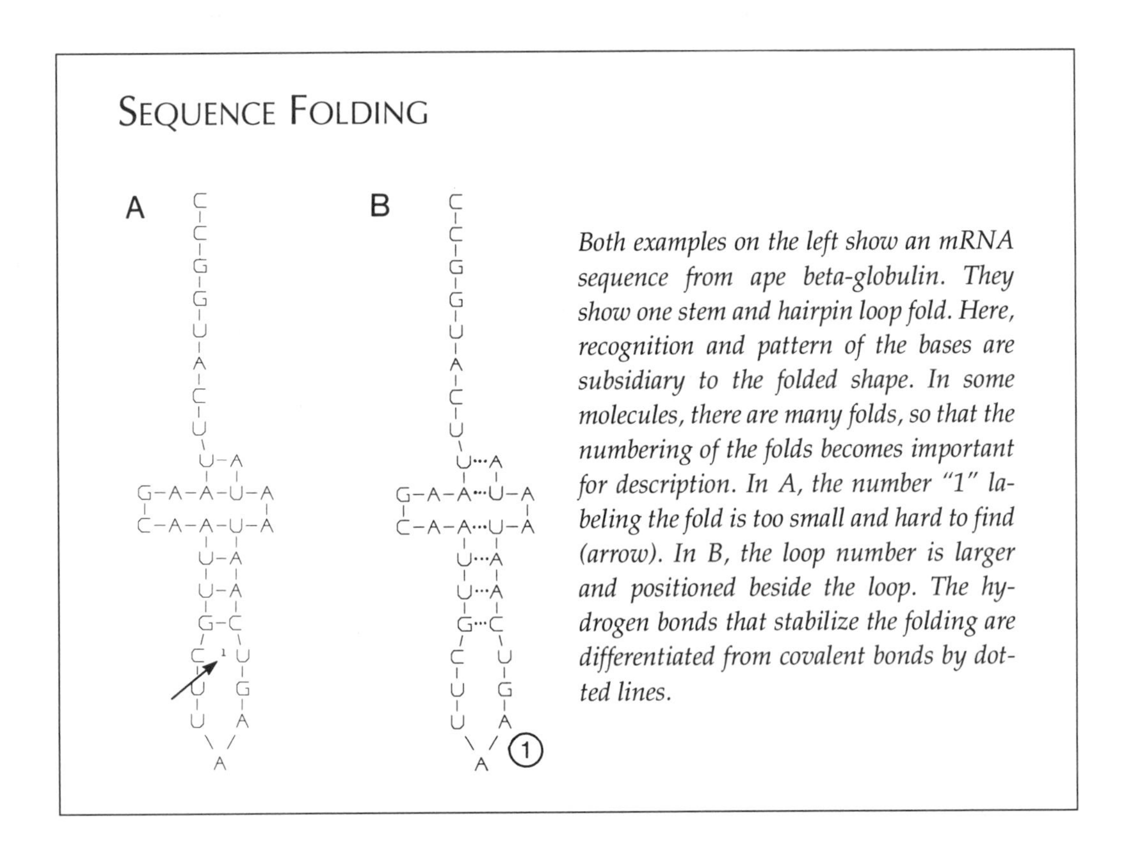

SEQUENCE FOLDING

Both examples on the left show an mRNA sequence from ape beta-globulin. They show one stem and hairpin loop fold. Here, recognition and pattern of the bases are subsidiary to the folded shape. In some molecules, there are many folds, so that the numbering of the folds becomes important for description. In A, the number "1" labeling the fold is too small and hard to find (arrow). In B, the loop number is larger and positioned beside the loop. The hydrogen bonds that stabilize the folding are differentiated from covalent bonds by dotted lines.

Folding of protein is shown in the upper figure on the next page.

Protein Structure Wheel

This simple model generated from an amino acid sequence determines and predicts the alpha helical and beta sheet regions in protein. The helical net on page 62 shows a flattened view of the alpha helical region of a protein. The wheel on the next page suggests a more helical arrangement.

PROTEIN FOLDING

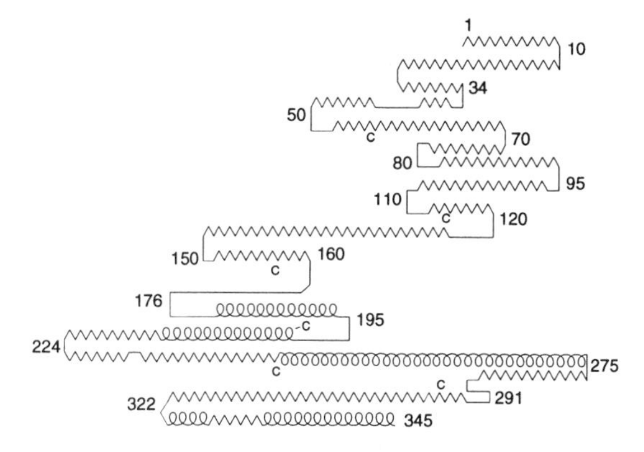

Folding of protein was predicted by the sequence method. Amino acid abbreviations are not shown here, and emphasis is on the beta sheets (sharp peaks) and alpha helices (loops).

PROTEIN WHEEL

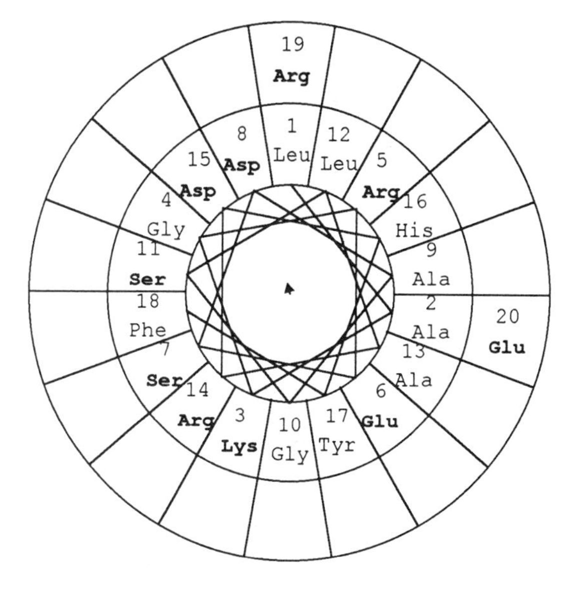

This wheel shows the alpha helical regions of a protein with heavy hydrophobic residues. This model was generated by the DNASIS *program. Protein structure wheels showing alpha helices are available in many programs.*

Hydrophobic residues
Hydrophilic residues
Mean hydrophobicity: 0.71
Hydrophobic moment : 0.08

Folding diagrams are limited to one plane, but the bonding and the rudimentary probability of the structure are shown. For an idea of tertiary structure, it is necessary to simulate a third dimension. The ribbon diagram below is one way to show three-dimensional folding in a protein, based on bonding, alpha helical structure, and beta strand structure.

RIBBON MODEL

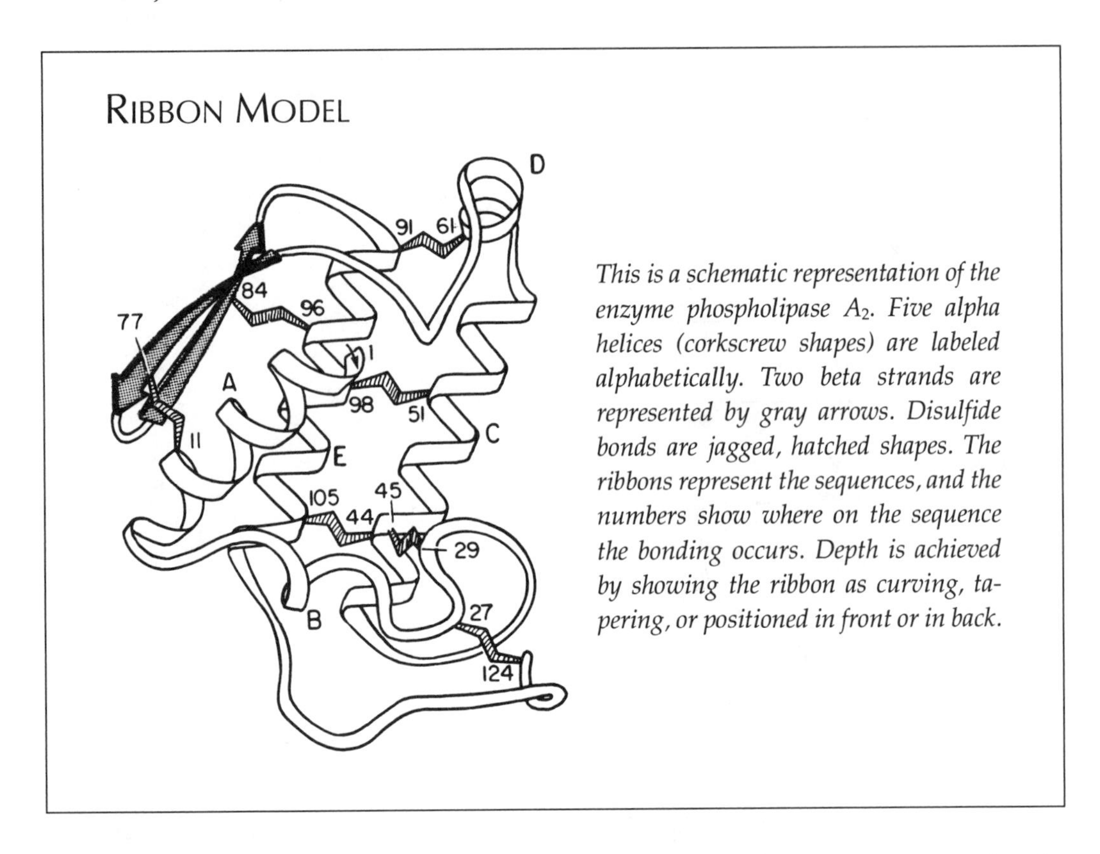

This is a schematic representation of the enzyme phospholipase A_2. Five alpha helices (corkscrew shapes) are labeled alphabetically. Two beta strands are represented by gray arrows. Disulfide bonds are jagged, hatched shapes. The ribbons represent the sequences, and the numbers show where on the sequence the bonding occurs. Depth is achieved by showing the ribbon as curving, tapering, or positioned in front or in back.

More conventional models show the bonding of atoms to form a molecule. These models are specified as skeletal, ball and stick, and space filling. There is a great deal of symbolism involved in the representation of these models: a circle for an atom and a stick for a bond, for instance. Most of this symbolism is commonly understood, and care should be taken to use it accurately or to explain carefully any deviations from common usage. Symbolism such as size differences and tapering lines is used to convey a sense of three dimensions.

Depth is suggested in the skeletal model by the angles of the lines, but for a more realistic, in-depth view of the structure, stereo models can be generated that are extraordinarily realistic.

The stick-and-dot model shows the skeletal structure of a molecule with the surrounding atoms ghosted in.

SKELETAL MODEL

A skeletal model is the simplest model. Only the bonds with their junctions and endings are shown. An attempt is made to position them in space by angling the bonds away from a plane and slanting the plane of the pentagon.

STEREO MODEL

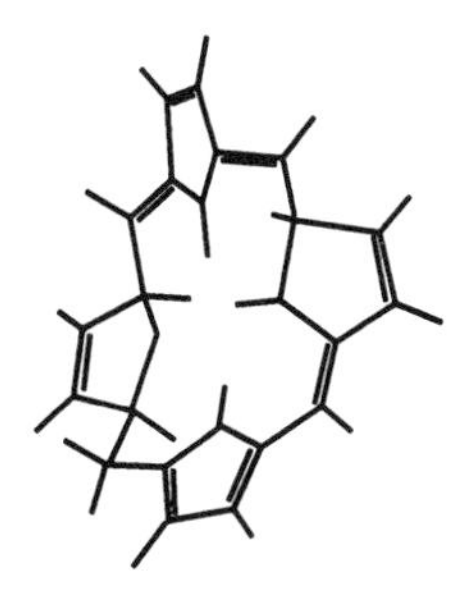

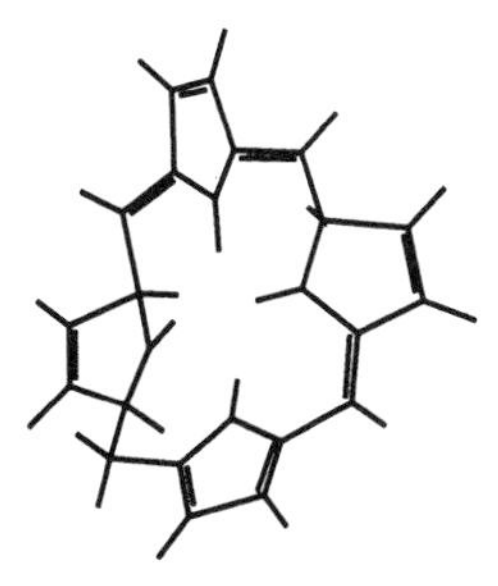

This stereo model of porphyrin was generated in HyperChem® release 4 by Hypercube Inc. and Adobe Illustrator™. By staring cross-eyed, or relaxing and staring at nothing, the two slightly offset figures come together to float in three-dimensional space.

BALL-AND-STICK MODEL

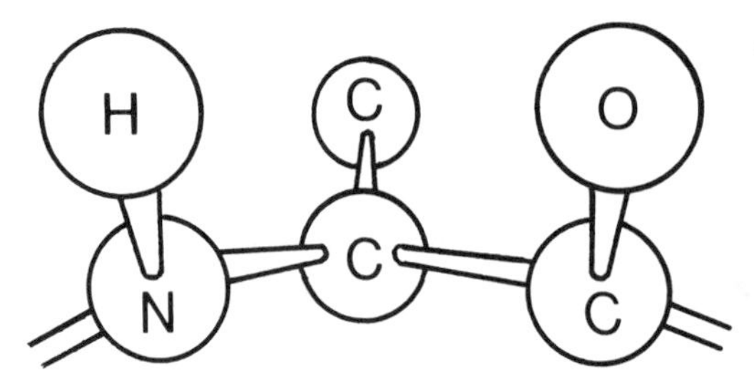

Here the ball represents the atom, the stick represents the bond. Depth in space is suggested by the tapering of the stick as well as by the stick's position in front or behind the atom. The size of the atom is also relative to its position in space. The three kinds of molecules are distinguished by labels, but for a slide, the atoms could easily be color keyed.

STICK-AND-DOT MODEL

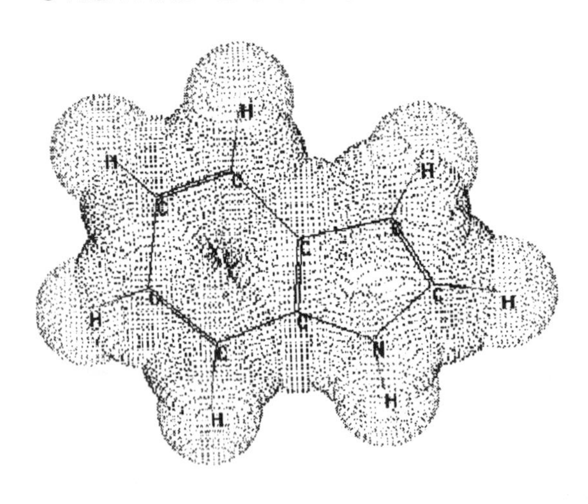

This model shows the molecule indole. Hydrogen and carbon bonds are labeled, and double bonds are shown. This was created in HyperChem® and Adobe Photoshop™ software.

This type of model is best shown in color to distinguish the kinds of molecules and to make the labels clearer. It is generally shown with jewel colors on a black background.

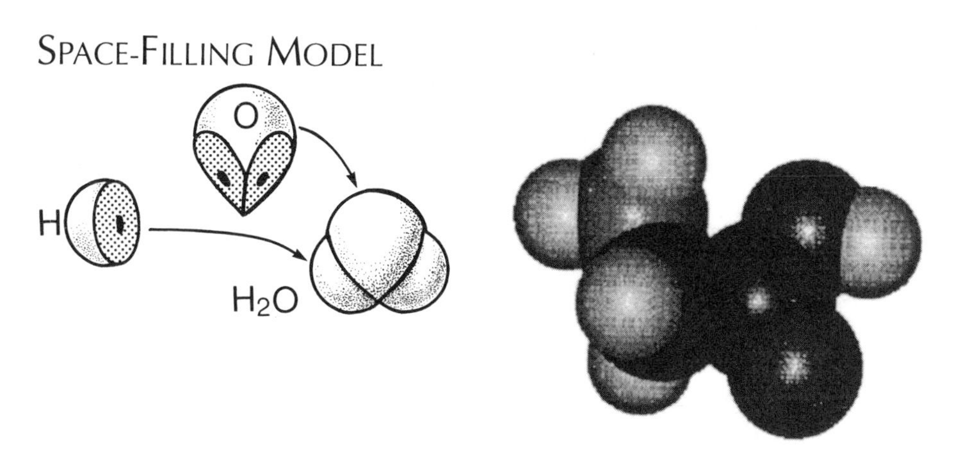

The model on the left was hand drawn and shows how the hypothetical shapes of the atoms could fit together to form a molecule. A three-dimensional effect is achieved by shading. The model on the right was generated by HyperChem® and Adobe Photoshop™ and shows the molecule alanine. This kind of model is not as effective in black and white as it is in color, which clarifies the difference in the atoms.

The space-filling model is more realistic than the skeletal or ball-and-stick models, because it shows clusters of atoms adhering closely. The bonds, which are theoretical constructs anyway, must be imagined.

All these kinds of models leave much to be desired in molecules that are much larger and more complex than have been shown in the above diagrams.

Model Shortcomings

- Inside atoms are hidden by outside atoms.
- It is difficult to distinguish the different components of a large molecule.
- It is impossible to see all aspects of a molecule's structure in a two-dimensional diagram.

The Computer Graphics Laboratory at the University of California, San Francisco, founded by Robert Langridge and now headed by Thomas Ferrin, has pioneered work in molecular modeling. This group has made great strides in overcoming the limitations of conventional models.

They have designed computer programs that show a buildup of the layers of atoms that comprise the space-filling model.

Bonds and clusters of adjoining molecules are easily distinguished by color in programs that the Computer Graphics Laboratory has designed.

All aspects of a molecule may be seen using a program that has been designed to rotate the molecule.

Other striking capabilities offered in their programs are

- Stereoscopic presentation of molecules show lifelike depth and dimension.
- Pocket sites for enzyme interaction are clearly color coded, and the interaction between two molecules is shown.
- Movement of proteins is shown with possible places for fit to other molecules.

The Computer Graphics Laboratory is connected to networks all over the world and is available for consultation. This group likes to put its program in the users' hands and will help to customize the program.

Such advanced technology offers tools to facilitate communication. However, some of these tools have outpaced the pedestrian media that are still being used: the journal, poster, and slide media. Stereoviews are being used in some journals today, but there may come a day when video tapes and computer disks will be mailed instead of journals, when movies, videotapes, and computer multi media will be used instead of slides and posters.

Meanwhile, there are still crucial decisions to make:

- What view of a three-dimensional structure should be used in a two-dimensional presentation?
- What will work best for the media available?
- How important is color to the presentation?
- How much money can be allotted for a given figure?

The *facet of the model* chosen for presentation should conform to the facts to be emphasized. For showing the three-dimensional structure, short of using a revolving model on the computer or TV screen, several sides of the model can be depicted in several different diagrams.

For any *medium* used, a photograph of the model generated on the computer screen, may be appropriate. A color slide, or an enlarged color print shot from the screen can be effective. A slide made from the computer screen, with its black background and vivid colors, can be spectacular. Although the colors for a poster print lose some brilliance, the resulting figure is still striking. For publication, the negative or print must go through another step in reproduction, with a probable loss in brilliance.

Color and Cost

Often a model must be shown in *color* to communicate information clearly. For this reason, journals are printing more color figures taken from the computer screen.

Indeed, they frequently provide an eye-catching journal cover. Color figures of this kind cannot be reproduced well in black and white. Meaning and distinctions are lost in the subtle grays that result.

The disadvantage of color prints for publication is their *cost.* Although slides and prints of this kind are not unreasonable, because of the process of making a color print, the journal charge for color prints is quite high. If color is necessary to make clear distinctions, high cost may not be a deterrent. For work that is substantial in its findings and in the labor required for the result, the high cost may be justified.

If the findings require many figures or if grants are not large, consider using a black and white graphics program or an artist.

It takes time and thought to solve the problems of visual communication that continue to arise as new information is brought to light and new graphical tools are developed. However, as long as the desire to communicate coexists with a knowledge of principles of good visual communication, time and thought on this subject will be well spent.

SOFTWARE COMPARISON

The intention of this book is not to espouse or recommend any particular computer software for molecular graphics. There is a wealth of programs available, and new and better ones are on the market every day. Existing programs are constantly updated.

If you are in the market to buy a program, be very clear on your own priorities before you research the market. It is probable that you will want to communicate future work pictorially. So make graphical flexibility one of your requirements.

Examples of and differences in software have been illustrated in this chapter, and the programs used have been cited. Below is a list of the software shown in this chapter. Most manufacturers send literature, and some send demonstration disks.

Sequencing Software

GeneWorks®	IntelliGenetics, Inc. 700 E. El Camino Real Mountain View, CA 94040 Fax: (415) 962-7302; phone: (415) 962-7300
DNA Strider™	Christian Marck, Service de Biochimie—Bat. 142 Centre d'Etudes Nucleaires de Saclay 91191 Gif-sur-Yvette Cedex, France

DNASIS	Hitachi Software Engineering American Ltd. 1111 Bayhill Drive, Suite 395 San Bruno, CA 94066 Phone: (415) 615-9600
Lasergene	DNASTAR, Inc. 1228 South Park Street Madison, WI 53715 Fax: (608) 258-7439; phone: (608) 258-7420

Molecular Modeling Software

HyperChem® release 4	Hypercube 419 Phillip Street Waterloo, Ontario, Canada N2L3X2 Fax: (519) 725-5193; phone: (519)725-4040 or (800) 960-1871

Most computer programs are designed for analysis, not for good graphic communication. Look for a program that has graphical as well as analytic capabilities. Or look for a program that allows transfer to drawing programs for changes and additions.

6
GRAPHS AND SOFTWARE

A graph is a powerful way to show the results of mathematical calculation. Historically, graphs are a recent invention, a product of the scientific revolution of the seventeenth century. In *La Géométrie,* published in 1679, René Descartes arithmetized geometry by fixing the positions of points on a plane by the use of coordinate axis lines.

The concept of showing change and processes by graphing has undoubtedly played a crucial part in the development and understanding of science. Isaac Newton was the first to understand that change could be expressed graphically, although he fell back on geometrical forms in his *Principia* (possibly one reason it is so difficult to read). To describe changes, relationships, and trends in words alone is a lengthy and complicated task and, once done, may still be hard to understand.

If you have a choice of presenting your information in tables or graphs, choose the graph. A graph conveys the information more quickly and easily than a table. It also shows the information more impressively and memorably.

However, if the information can be said in one or two sentences or if the absolute numerical values are necessary in the presentation, use words or tables. To emphasize essential numerical data, use a graph *with* the table.

A graph is a system of connections expressed by means of commonly accepted symbols. As such, the symbols and symbolic forms used in making graphs are significant. To communicate clearly this symbolism must be acknowledged.

Software for computer graphs uses accepted symbolism but does not automatically produce good graphs. Raw computer-generated graphs (or default graphs) require thought and time to evolve into graphs that communicate clearly and effectively. The software on the market now offers more options than you will usually use. But none of the programs are perfect, and none produce good graphs without effort and experimentation.

This chapter shows examples of problems of confusion and ambiguity encountered in graphs and suggests solutions. In this chapter kinds of graphs and design requirements are discussed and illustrated by several graphing programs.

The commonly used graph forms are the pie, bar and line graphs. Each shows data differently. Computer graphing programs offer options for all of these kinds of graphs along with other lesser used forms. It is also possible to make a variety of kinds of graphs from one set of data. Below are examples of frequently used graphs available in graphing software.

KINDS OF GRAPHS

The Pie Graph

The symbol of the circle or the "pie" with its wedge-shaped pieces is symbolically satisfying and easily understood by the viewer. It shows the whole with its parts.

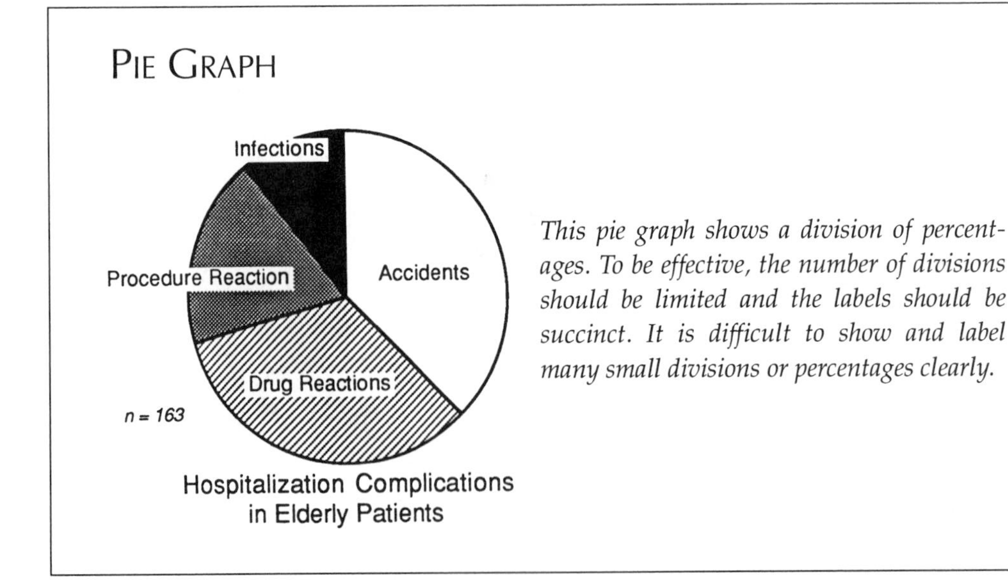

This pie graph shows a division of percentages. To be effective, the number of divisions should be limited and the labels should be succinct. It is difficult to show and label many small divisions or percentages clearly.

The Bar Graph

The symbolism of the bar graph is as familiar to us as the yardstick or the column of mercury in the thermometer. It shows measured amounts. The simplest bar graph compares one or more sets of measurements on a single axis.

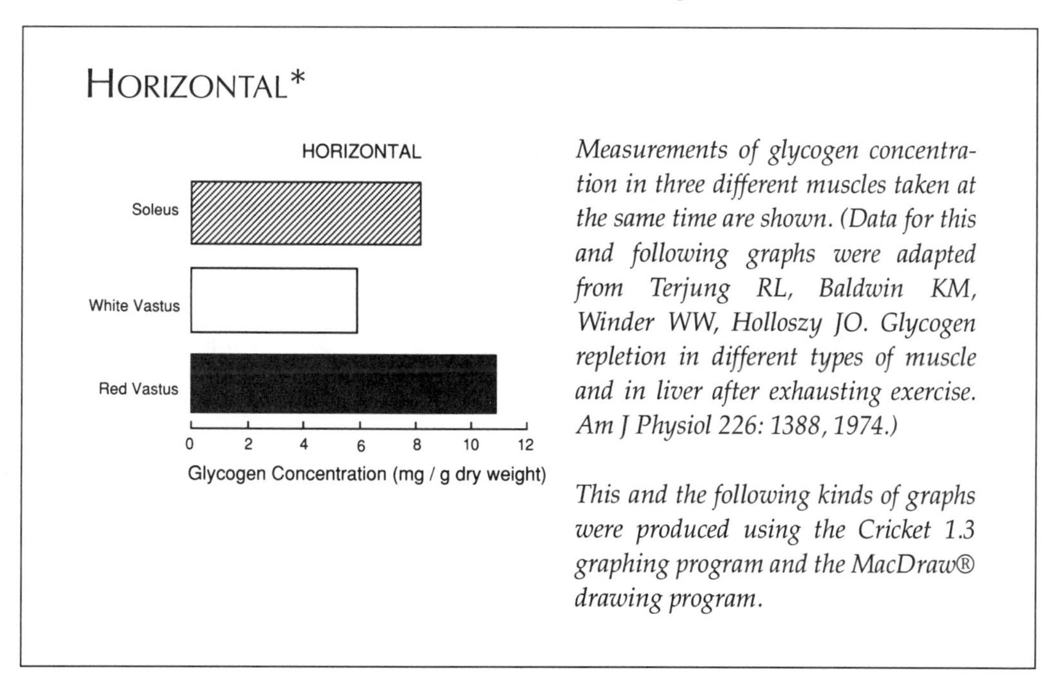

HORIZONTAL*

Measurements of glycogen concentration in three different muscles taken at the same time are shown. (Data for this and following graphs were adapted from Terjung RL, Baldwin KM, Winder WW, Holloszy JO. Glycogen repletion in different types of muscle and in liver after exhausting exercise. Am J Physiol 226: 1388, 1974.)

This and the following kinds of graphs were produced using the Cricket 1.3 graphing program and the MacDraw® drawing program.

Horizontally arranged bar graphs, as in the previous graph, are more commonly used in business than in science, but for writing long labels horizontally and without abbreviations this arrangement is clearly practical.

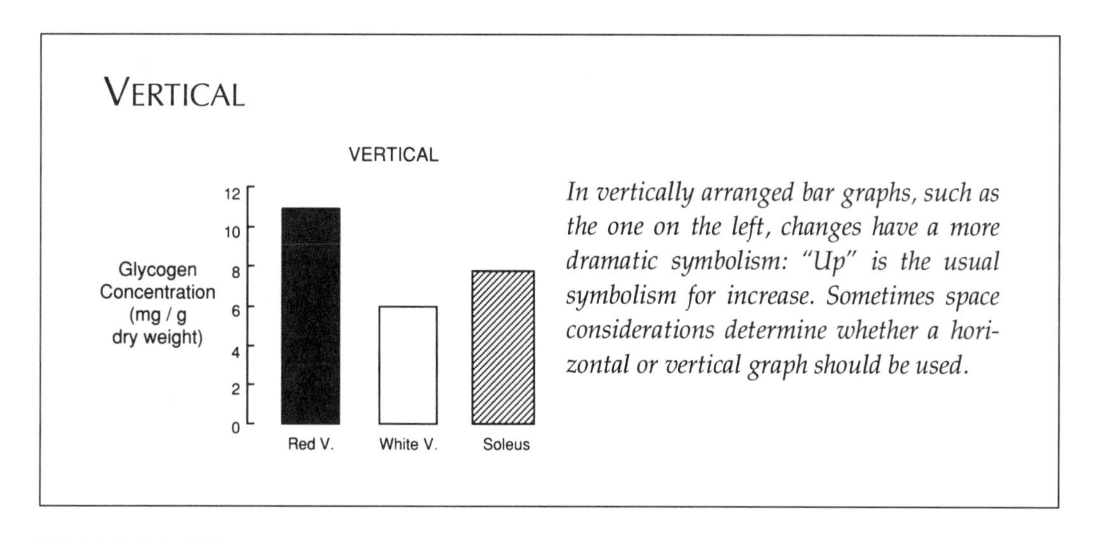

VERTICAL

In vertically arranged bar graphs, such as the one on the left, changes have a more dramatic symbolism: "Up" is the usual symbolism for increase. Sometimes space considerations determine whether a horizontal or vertical graph should be used.

*Graphing programs usually call the horizontal arrangement a "bar graph" and the vertical arrangement a "column graph."

In bar graphs, measurements made at different times may be shown. However, such measurements are static or unchanging. Bar graphs show separate measurements by themselves or compared with other separate measurements.

MULTIPLE COMPARISONS

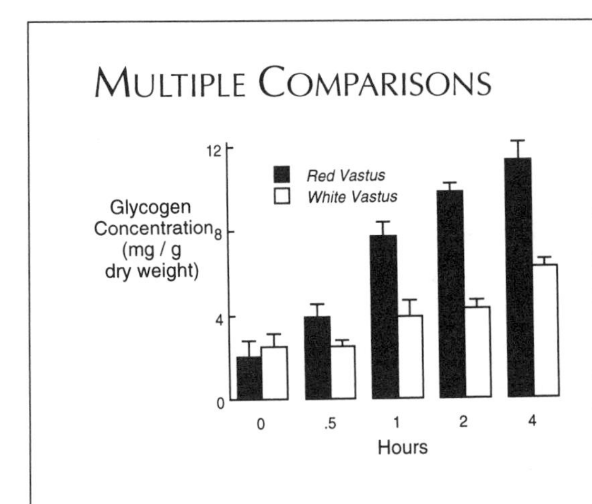

The bar graph to the left compares two sets of data measured at five different times. The red and white vastus are distinguished by shading and labeled in the key. There is no horizontal axis line because there is no second scale.

TOTALS AND COMPONENTS

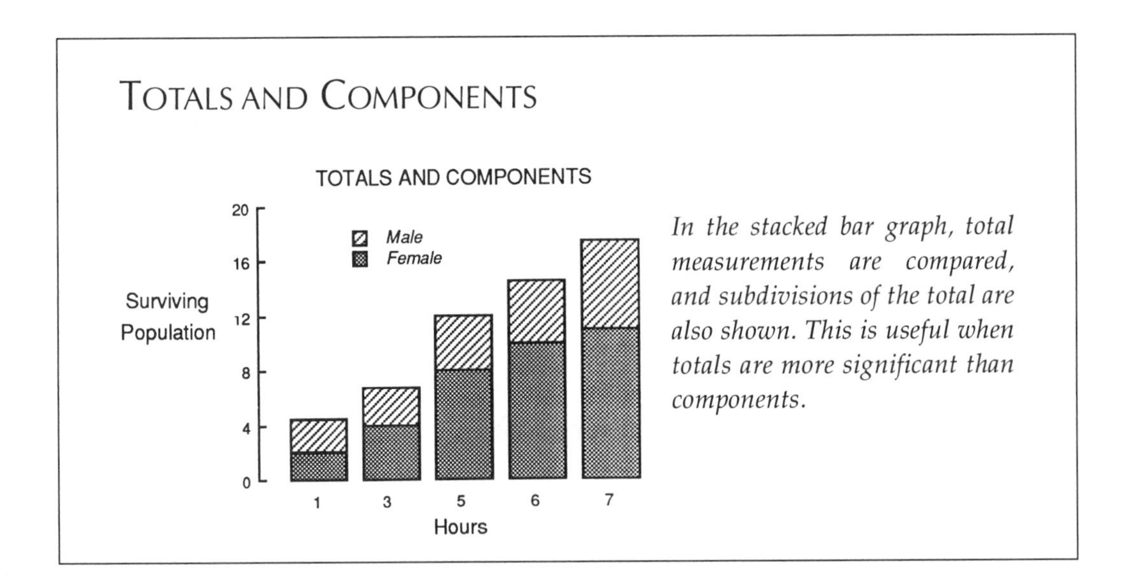

In the stacked bar graph, total measurements are compared, and subdivisions of the total are also shown. This is useful when totals are more significant than components.

Many graphing programs offer options for three-dimensional bars or shadowed bars. This not only misrepresents the information because the information is one-dimensional but the devise itself becomes an obtrusive element that distracts the viewer. Avoid "special effects" if they do not enhance the point to be made.

The Histogram

A histogram measures frequency of occurrence. It could be expressed as a series of bars but usually shows only the tops of the bars.

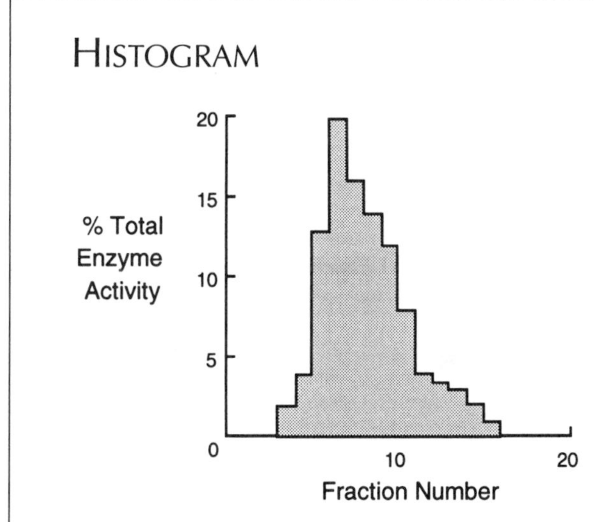

The graph on the left shows only the tops of the bars. It shows an equilibrium density distribution pattern and is a true histogram. The width of each bar represents a range of values. The shading is optional but emphasizes the area from baseline to the tops of the bars.

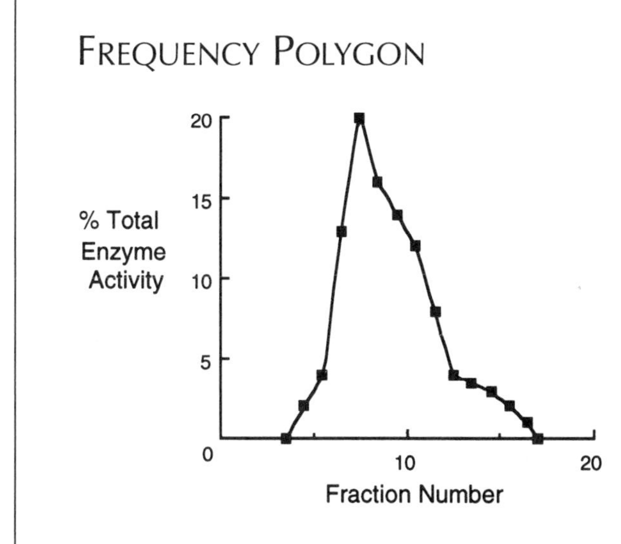

This is the same graph as above, but the frequency is plotted at the midpoint of the bars and joined by interpolated lines. This is referred to as a frequency polygon. The emphasis here is on the flow of change.

The Line Graph (or Time Line Graph)

Descartes's fixing of a point's position on a plane by use of coordinate axis lines was naturally extended by Newton to express movement. Change is shown when the points are connected by lines. The points and lines move the eye, and like the towns and highways symbolized on a map, they simplify and clarify large amounts of information.

LINE GRAPH

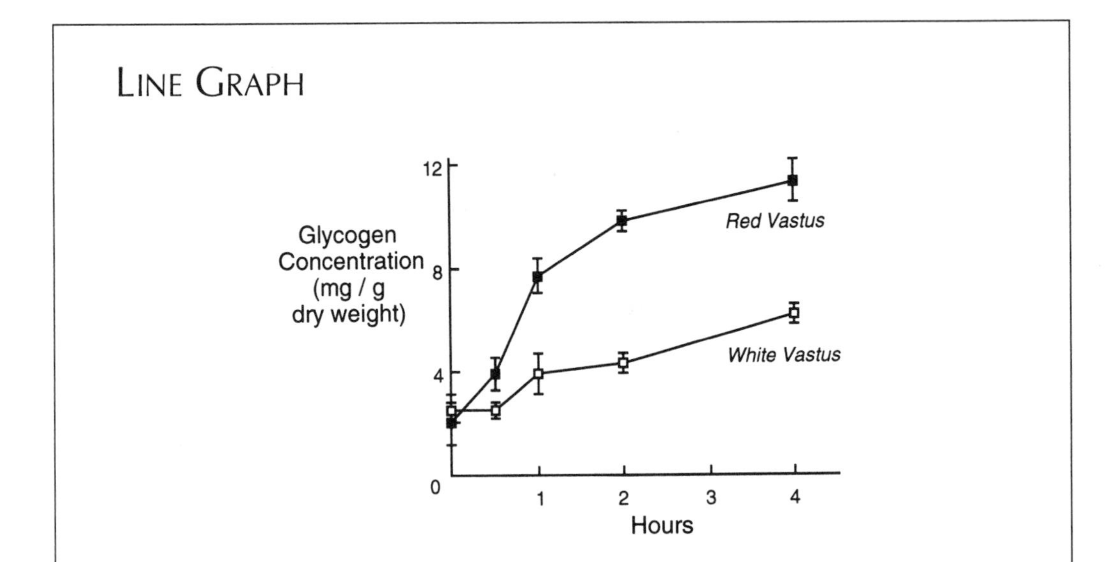

The points here represent the means of a quantity of collected data; the standard error lines show the spread. The lines joining the points create the impression of movement showing the change, trend, or comparison.

The Scattergram

The scattergram shows the influence of one variable on another. It shows the functional relationship of the variables. Points that are grouped together indicate strong correlation whereas points that are widely scattered indicate a weak relationship.

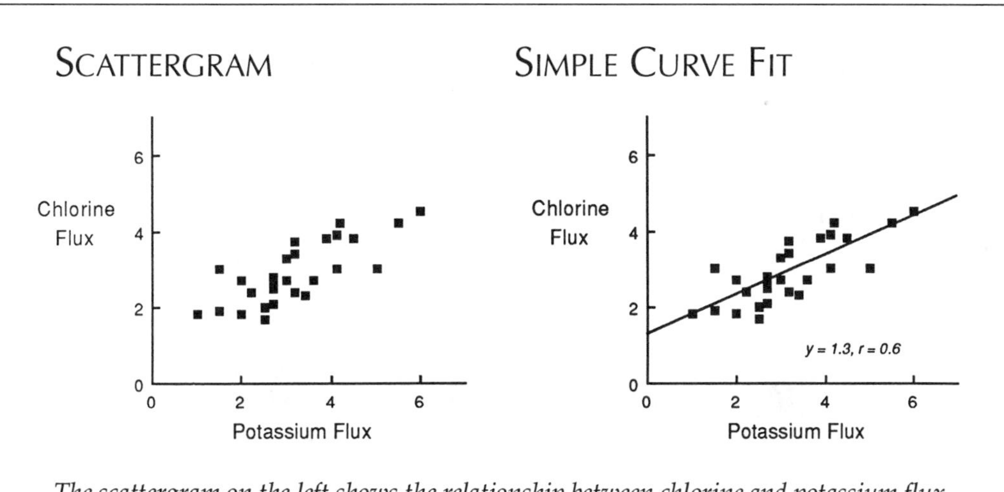

The scattergram on the left shows the relationship between chlorine and potassium flux. By adding the simple curve fit, as on the right, the correlation, or lack of correlation, becomes clearer. The correlation coefficient, r, *was calculated and plotted by the Cricket Graph™ program.*

Overlapping Graph Forms

Sometimes the information may fit several graph forms. Knowing the distinctions between the kinds of graphs and being clear about what should be emphasized will help in deciding which form of graph will work best.

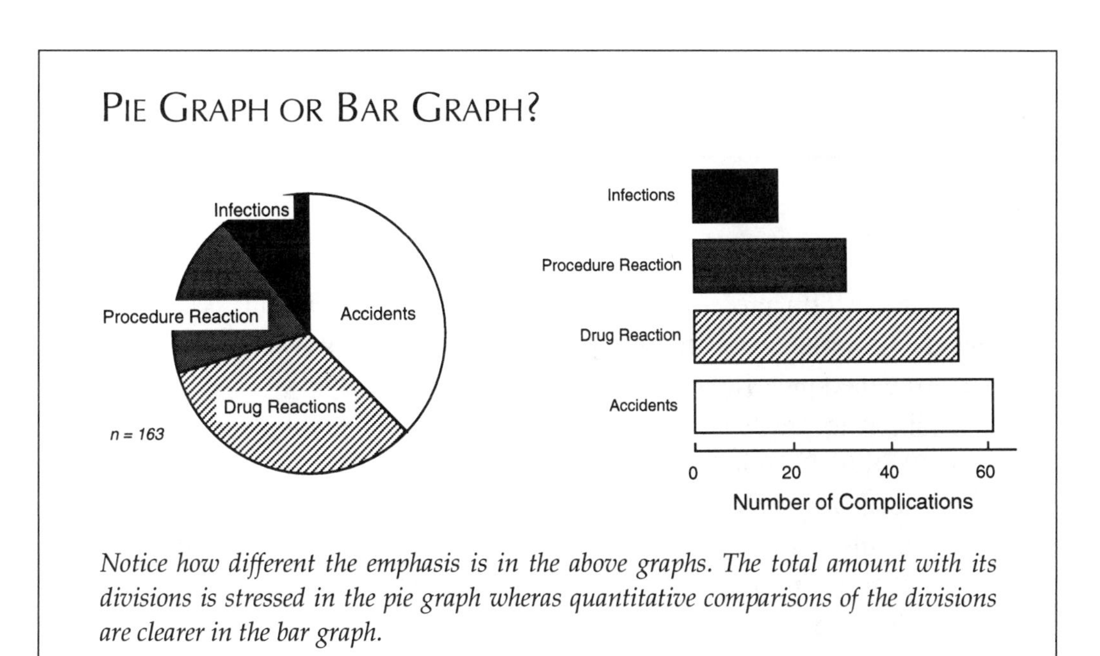

Notice how different the emphasis is in the above graphs. The total amount with its divisions is stressed in the pie graph wheras quantitative comparisons of the divisions are clearer in the bar graph.

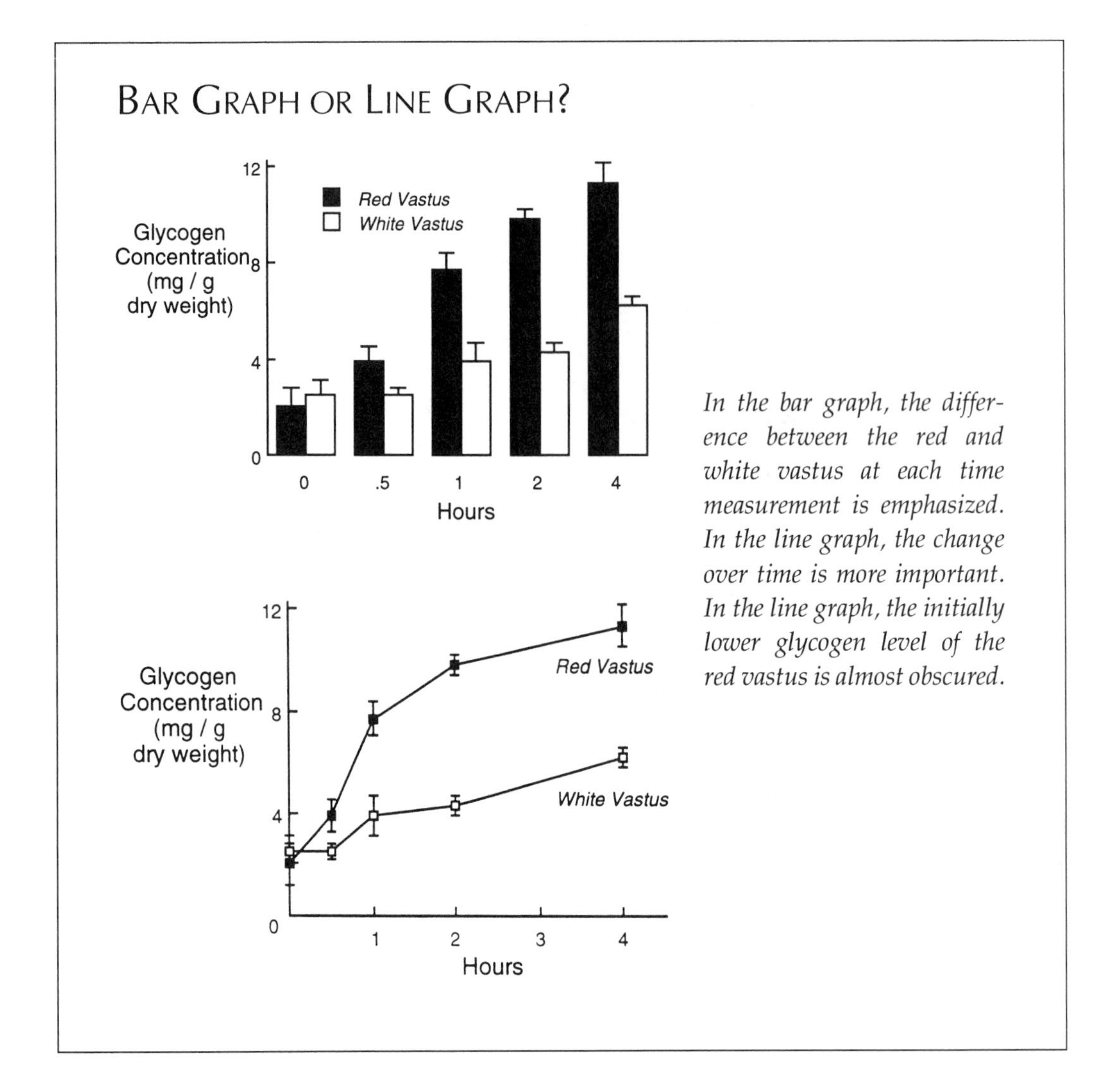

BAR GRAPH OR LINE GRAPH?

In the bar graph, the difference between the red and white vastus at each time measurement is emphasized. In the line graph, the change over time is more important. In the line graph, the initially lower glycogen level of the red vastus is almost obscured.

Each graph form presents information in a special way and uses particular symbolism to communicate an appropriate message. Know what you want to say and make sure the kind of graph you select fits your message.

In addition to choosing the appropriate form, the graph must be carefully tailored to communicate information forcefully. Graphing software offers most of the options needed to design excellent graphs and the criteria for good graph design have to do mainly with axes, labels and symbols. Unusual or experimental graphs that do not fit software parameters can often be started in a graphing program and refined in a drawing program.

GRAPH DESIGN

A good graph is uncluttered, clear, and focused. The graph below is cluttered, confusing, and focused on the wrong thing. It was produced in the automatic or default setting of a graphing program.

POOR GRAPH

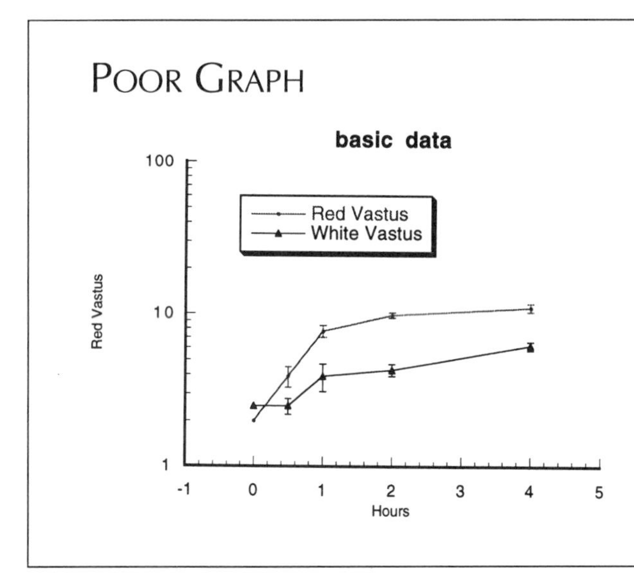

This is the raw or default graph generated by Kaleidagraph™ 3.0.2. It shows glycogen replenishment in two kinds of muscle after exercise. Heavy axis lines, the log scale, the boxed legend, too many ticks, unclear symbols, and small and misleading labels make this an unfocused and unclear graph.

IMPROVED GRAPH

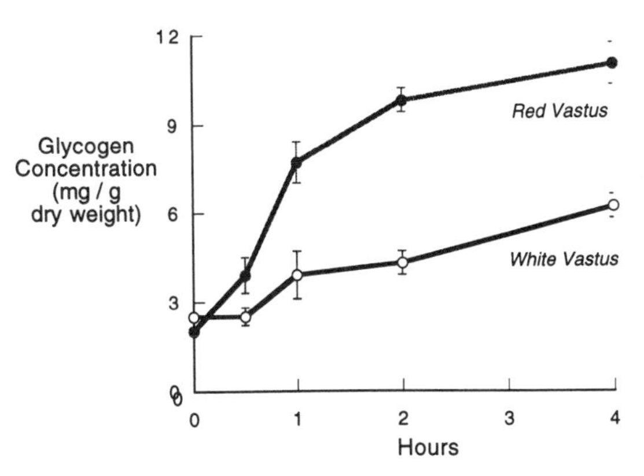

The y-axis is linear, ticks are fewer and regularly spaced, weight of lines is proportional to their importance, symbols are larger, and labels are added. The focus is now on the shapes of the muscle curves, emphasizing the important information. Note that in Kaleidagraph™ 3.0.2, the error bars and half of the error bar capping lines do not show up at 4 hours. Although this could be remedied by extending the x-axis to 5, wasted space would result. The alternative is to touch it up in a drawing program or by hand.

AXIS LINES

Most problems with graphs arise from misuse of axes: too heavy, too long, wrong intersection, ambiguous breaks, too many or too confusing increments, and incorrect proportions. What to look out for and how to correct these faults are discussed below.

An axis is the ruler that establishes regular intervals for measuring information. Because it is such a widely accepted convention, it is often taken for granted and its importance overlooked. Axes may emphasize, diminish, distort, simplify, or clutter the information. They must be used carefully and accurately.

The *x*-axis (horizontal axis, or abscissa) runs from left to right. The *y*-axis (vertical axis, or ordinate) runs from bottom to top.

Axis Length

To avoid wasting space, an axis should extend only to the labeled unit or minor tick closest to the last data point.

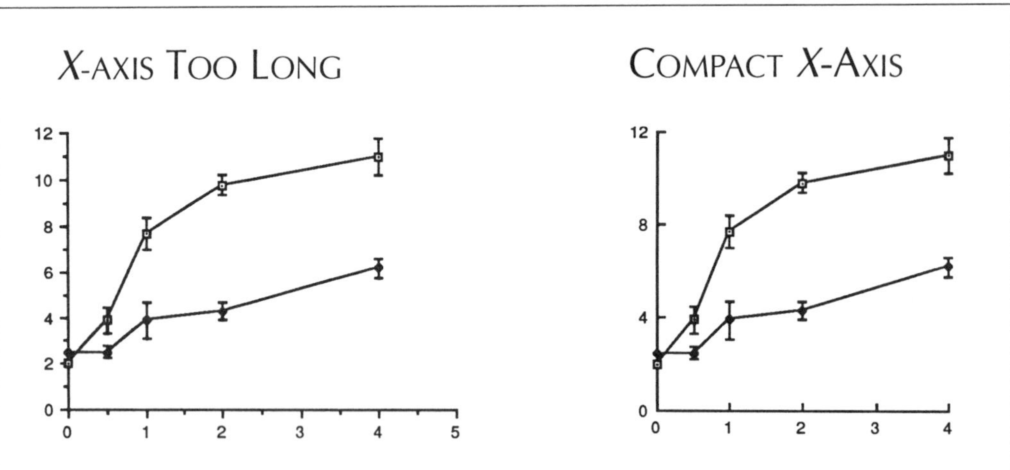

By ending the x*-axis at the last data point (4 hours), the graph on the right is more compact. In saving space, reduction in a journal is less and enlargement in a slide is greater. Most graphing software allows the shape of the graph to be shortened or lengthened so that a graph may be quickly and easily changed from slide format to publication format. This graph was produced in Cricket™ software. (Note that the error bars at 4 hours are intact.)*

Single Axis

In a bar graph, only one variable is presented. The second line is a base line that does not express anything. Below, a bar graph is shown in the KaleidaGraph™ program with and without the second line.

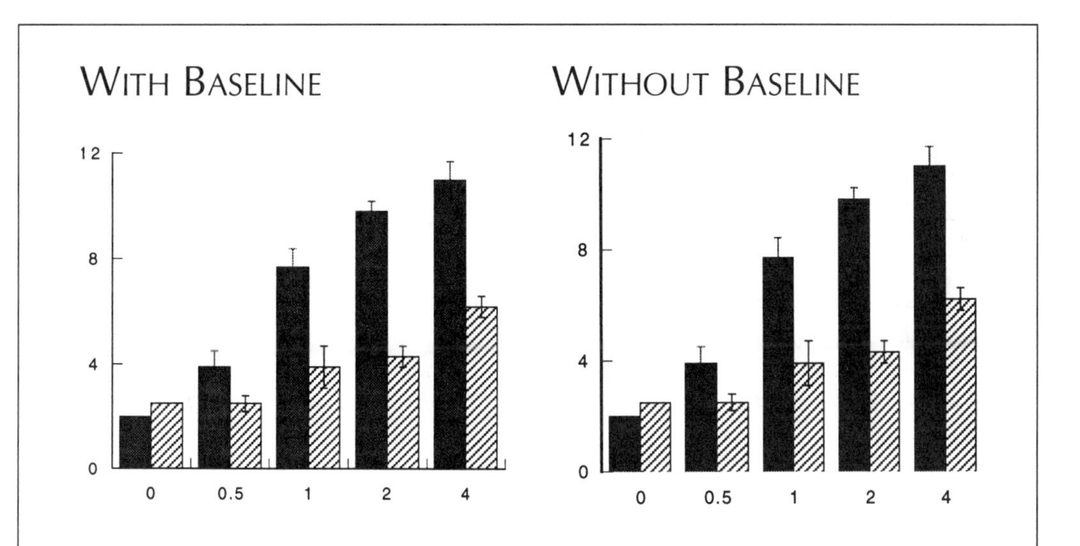

In graphing software, the bar graph, such as the one on the left, shows two axes. Because only the y-axis presents scaled data and because the measurements on the x-axis do not fall on a scale, the horizontal baseline should be left out. This is shown in the graph on the right. In addition to having a cleaner and more open look, the grouped measurements are more clearly shown to be separate, unscaled entities.

The KaleidaGraph™ program will allow axes removal, but in the case of the column graph, the bottom lines of the columns are also removed. (See the right-hand figure above.) Most computer graphic programs have no option for axis line removal. The superfluous line can be removed by transferring it to a drawing program. It can also be eliminated by white paint. In the KaleidaGraph™ program, the drawing tool can be used to add lines to the bottoms of the column.

Intersecting Axes

A line graph has an axis for each set of coordinates. The axes are two intersecting straight lines against which the relative positions of points are determined. The axes generally intersect at zero. If the two axes do not meet at zero, this fact should be indicated in a noticeable way.

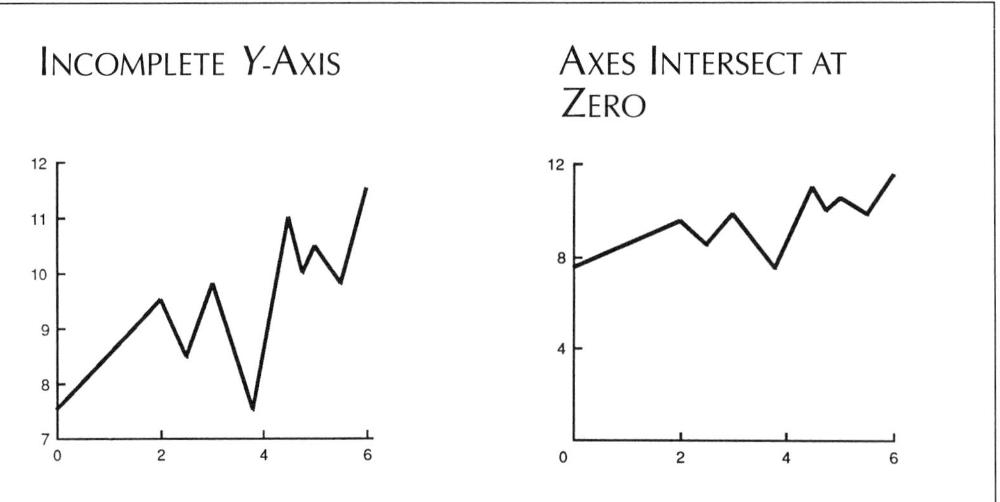

In the graph on the left, the x*-axis meets the* y*-axis at 7. The viewer must look twice to see that the* y*-axis is not complete. On the right, the whole* y*-axis is shown so that both axes meet at zero. This dramatically changes the look of the curve, and although it leaves unused space at the bottom of the graph, it may give the truest picture of the data.*

Starting an axis at zero may waste space at the bottom of the graph if data are limited to an area far above zero, as in the graph above. However, when comparing such a curve with others in a different y-axis range, comparisons are much clearer visually when all axes originate from zero. If changes in the curve are more important than where the curve falls on the axis, it is acceptable to start the axis at a number other than zero, as long as it is clearly indicated.

Indeed, sometimes there are valid reasons for not starting an axis from zero. Certain measurements do not start at zero (pH or logarithmic scales, for example), zero is irrelevant to some scales (wavelength or absorbance, for instance), or scales start below zero. Also, information may be obscured by an axis line. But whatever the reason for not beginning an axis line on zero, it should be made visually clear that this is the case.

One way to show that the axes do not start at zero is to offset the axes.

Unfortunately, most programs do not offer options for showing breaks or offset axes. The graph shown on the left at the top of the following page was done in the Cricket Graph™ program, then transferred to MacDraw® for changes. The graph on the right was done entirely in the Sigma Plot® program.

AXES OFFSET

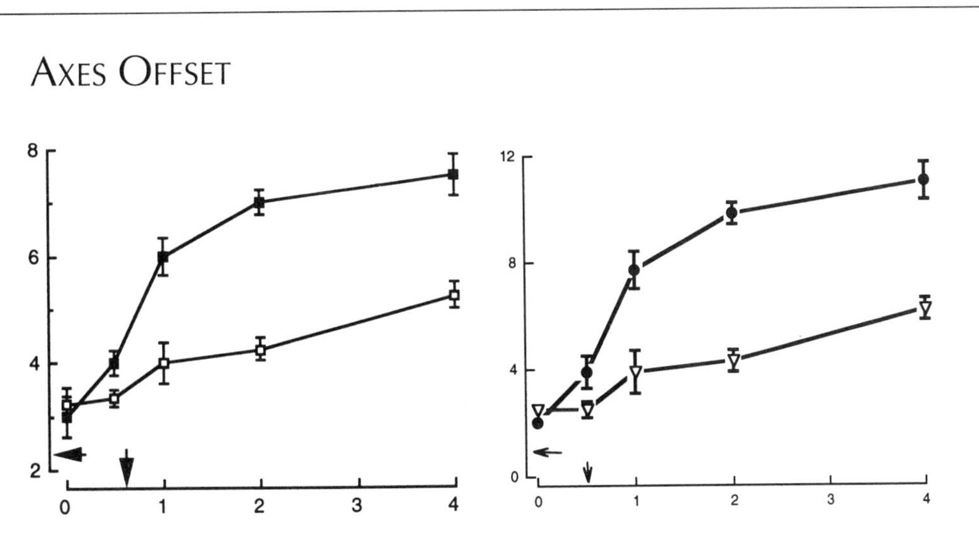

In the figure on the left, the y-axis starts at 2, but both axes have been moved aside or offset from their original position (arrows). It is immediately clear that the scales do not meet at zero. On the right, the axes have been offset (arrows) to make the points originating at zero clearer.

AXIS BREAK

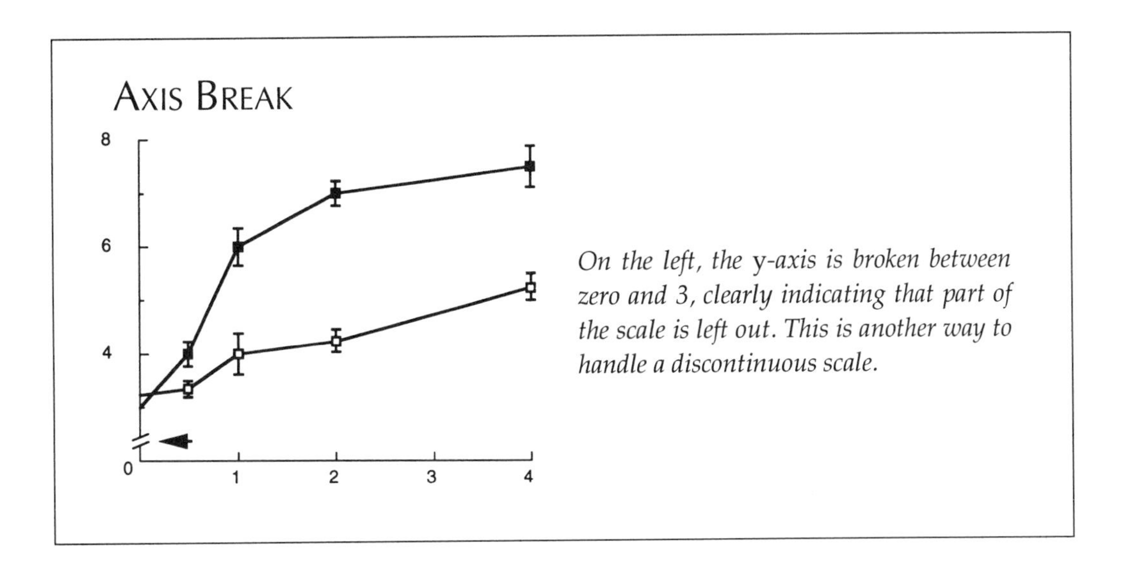

On the left, the y-axis is broken between zero and 3, clearly indicating that part of the scale is left out. This is another way to handle a discontinuous scale.

A scale break shows an interruption of continuous units of measurements. It is not accurate to use a scale break to show unit changes.

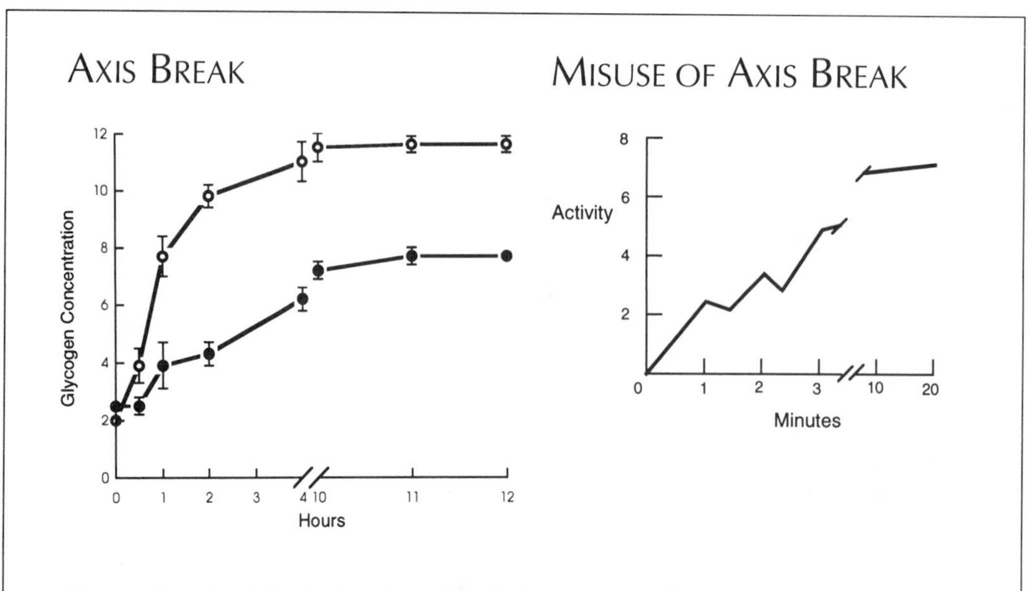

The x-*axis scale of the left-hand graph is broken at 4 hours and continues at 10 hours, indicating a lapse of 6 hours. From 10 to 12, the increments are by ones as they were before the break. On the right, increments before the break are by ones and after the break by tens. This violates the symbolism of a continuous axis line with regular intervals and may give a false first impression to the viewer.*

The graph on the left was plotted in Sigma Plot®, one of the few programs to have an axis break option. However, the length of the increments after the break is larger than before the break. This gives the impression of changed increments and is not logical. The graph on the right was done in the Cricket Graph™ and MacDraw® programs.

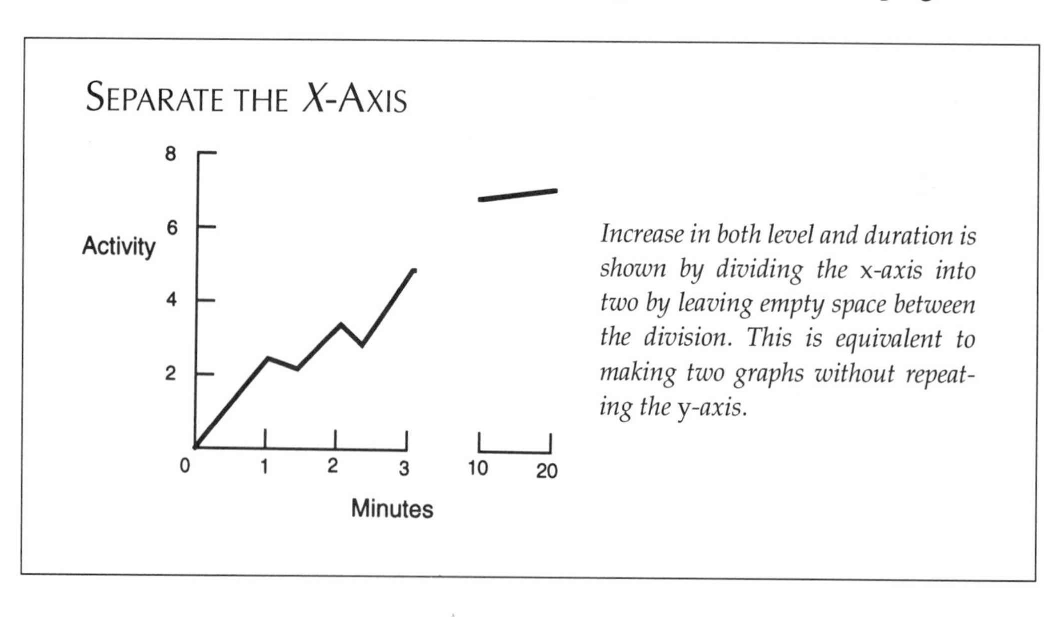

Increase in both level and duration is shown by dividing the x-*axis into two by leaving empty space between the division. This is equivalent to making two graphs without repeating the* y-*axis.*

If the difference in duration as well as level of activity after the axis break is important, completely separate the x-axis into two parts. (See the bottom figure on the opposite page.)

TICKS

Tick marks symbolize abbreviated grid lines. To show a whole grid is not desirable because it obscures curves and causes visual clutter. The number of tick marks and labels should be as even as possible for both axes. Crowding axes with ticks and numbers clutters the graph, lessening viewer comprehension. Also, one axis with crowded ticks and numbers draws attention to itself. Because ticks represent grid lines, they are better facing into the graph.

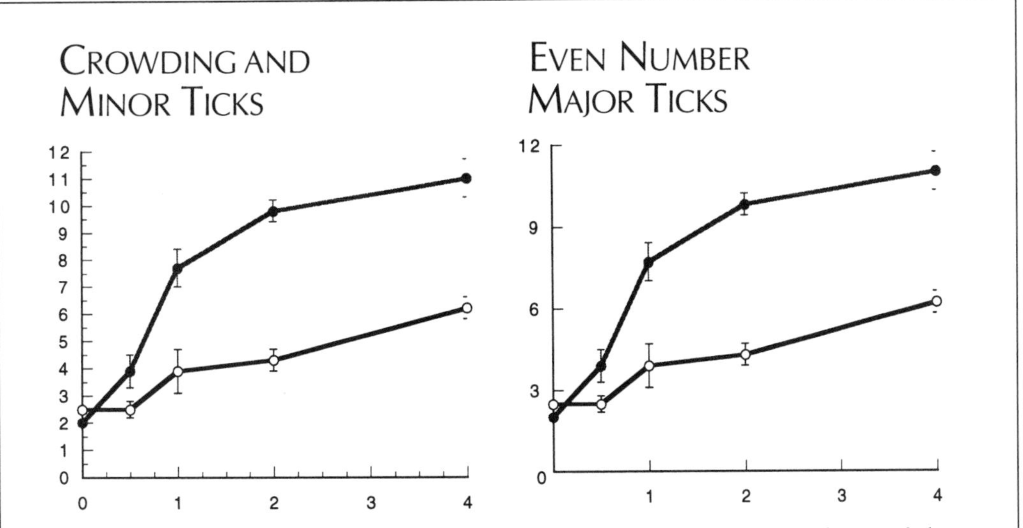

Cluttered scales with too many ticks and labels should be simplified. Also, if one scale has many more ticks than the other, as in the left-hand graph, attention is drawn to the crowded axis. The impression is that there is something different or important expressed by the crowding. The graph on the right without minor ticks is simple and uncluttered. The minor ticks serve no function.

The top graph on the next page has ticks that are divisible by 3.27. This is useless to the viewer. Increments should be easily divided.

Eliminate unnecessary zeros. This is also a cause of clutter. (See the bottom graph on the next page.)

An axis may be scaled in arithmetic units (linear scale) or logarithmic units (log scale). In the linear scale, as we have seen from the previous graphs, all

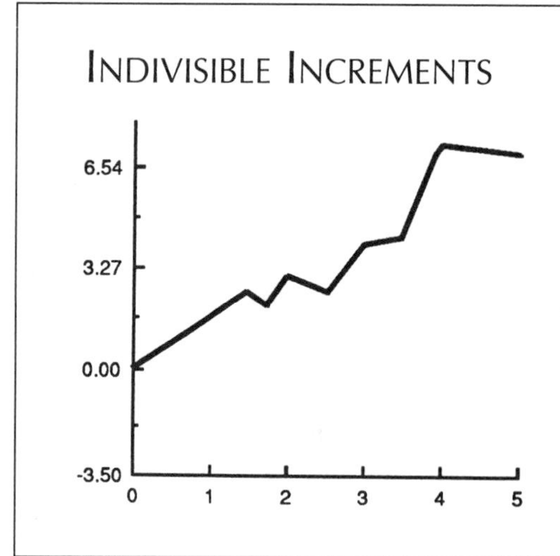

The major increments on the y-axis here are multiples of 3.27. Many graphs have odd increments because they do not start from zero or because of the limitations of a graphing program. Avoid software that has no option for adjusting increments.

increments are uniform and start from zero. In a log scale, units grow increasingly smaller for each cycle. For a scale of many log cycles, only the first number of the cycle need be labeled. Because the tick intervals are irregular on log scales, enough ticks should be included to show immediately that the scale is logarithmic.

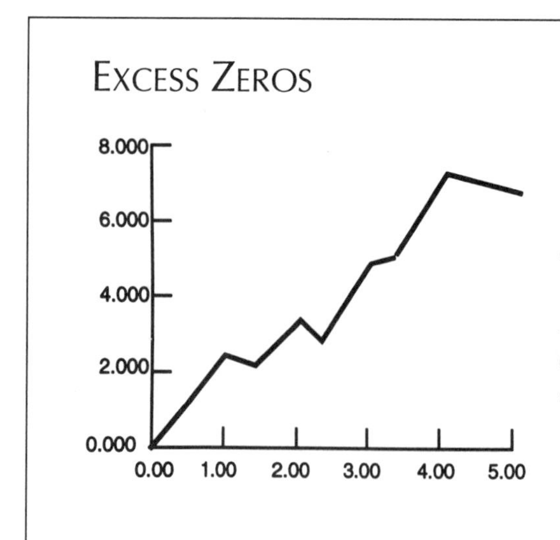

The zeros in the decimal places here serve no function and should be eliminated to avoid crowding.

If a graphing program shows excess zeros as a default, it should have a mechanism to change this.

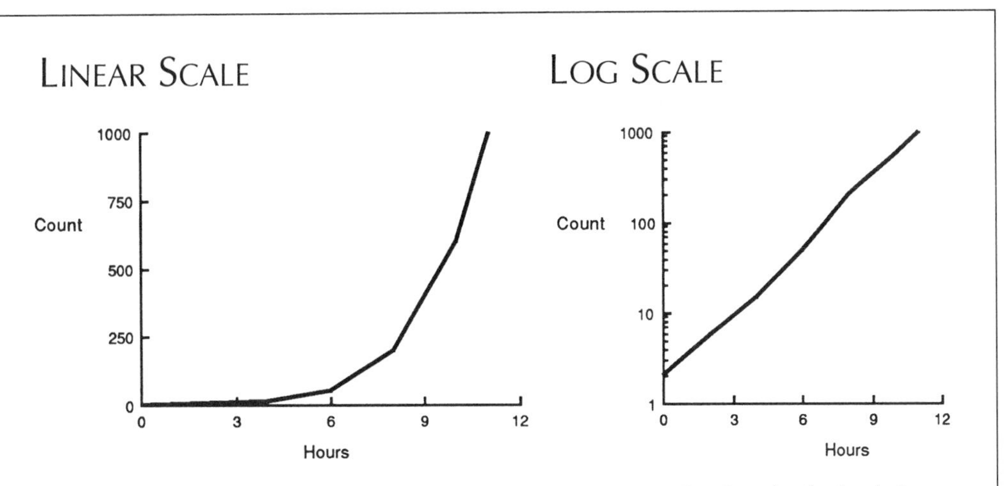

The ticks on the log scale are visual cues to the axis type. The data for both of these examples are the same. The log scale shows exponential increase, and the resulting curve looks quite different from that of the linear graph. The range of data and aspects of the data that are to be emphasized will determine which scale to use.

Axis Proportions

In general, axes should not extend beyond the last data point except to end the axis with a labeled tick. Both axes should be even in length as well as in number of units. However, adjustments in axis lengths, are sometimes justified to fit the space requirements of the medium you use or to express the data more truthfully.

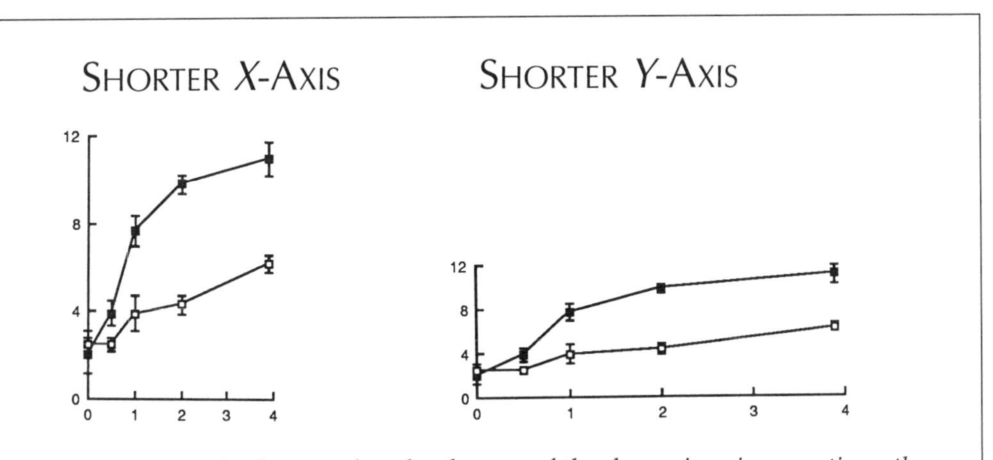

Both graphs contain the same data, but because of the change in axis proportions, the impressions from each are different. By shortening the x-axis, the curve increase is emphasized. By shortening the y-axis, the curve is flattened. Notice that the standard errors on the graph on the right make the data seem much more uniform.

By shortening the x-axis, labels may be written horizontally without increasing the reduction in a journal. If the similarity of increase of both curves is important, the shorter y-axis shows a truer picture. Also, if a wider format is desirable (for a slide for instance), a shorter y-axis works well.

Duplicate Axes

Duplicate axes are axis lines repeated at the top and right of the graph. This is referred to as "boxing in the graph" or "plot frame" in some graphing programs. These extra lines are unnecessary unless it is important to be able to make measurements across the graph. In that case, a table or grid might serve the purpose better.

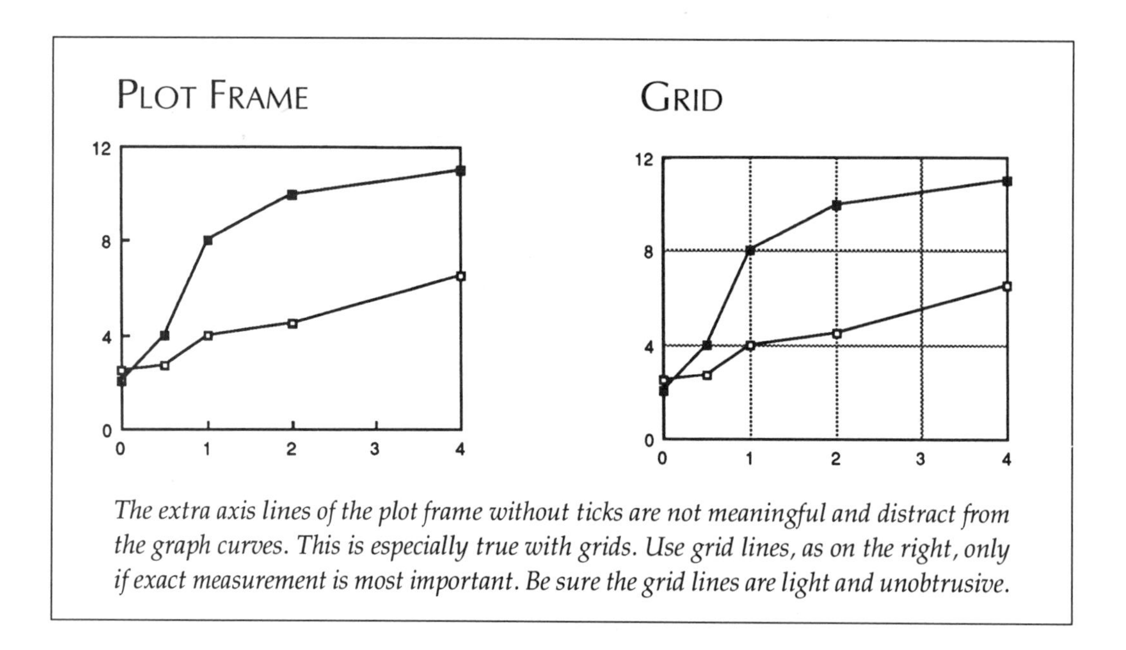

The extra axis lines of the plot frame without ticks are not meaningful and distract from the graph curves. This is especially true with grids. Use grid lines, as on the right, only if exact measurement is most important. Be sure the grid lines are light and unobtrusive.

> Computer programs make it easy to add extra lines. These lines, however, serve a dubious decorative function. Far from beautifying or compensating for paucity of information, they usually clutter or confuse.

LABELS

Labels should be complete but succinct. Long and complicated labels will defeat the viewer and therefore the purpose of the graph. Treat a label as a cue to jog the memory or to complete comprehension. Shorten long labels; avoid abbreviations unless they are universally understood; avoid repetition on the same graph. A title, for instance, should not repeat what is already in the axis labels. Be consistent in terminology.

The common problems in labeling are size, style, and placement. Although labels are necessary adjuncts for identification and explanation, they should not assume primary importance in the graph. This can happen when labels are too large or too bold or too long and complicated. However, labels must be large enough to be read easily.

In addition, label size, style, and placement may emphasize graph priorities. Variety in size and style may be functional as well as pleasing to look at, and the position of the labels may determine the sequence of attention.

Label Size

Label sizes should be in proportion to the size of the graph and to each other. Although all labels must be legible, they should not dominate the graph. Number size should be smaller than axis label size, because numbers are less important. The title should be larger than the axis labels because it is a general explanation that should be read first. Subsidiary explanatory labels such as the legend should be smaller than the axis labels.

The obvious problem of overly large or overly small labels is still commonly seen.

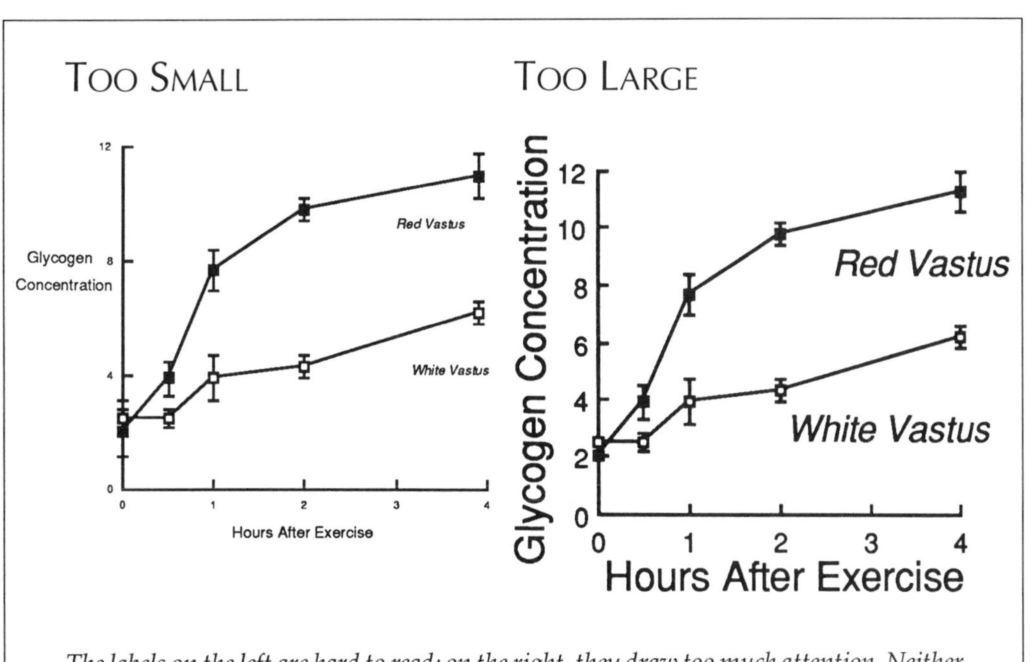

The labels on the left are hard to read; on the right, they draw too much attention. Neither satisfies us esthetically because the proportions are inappropriate. See the figures on the next page, "Upper Case Only" and "Lower Case Only," for optimum size labeling.

Label Style

Upper case letters and numbers are larger and more prominent than the same size lower case letters. The use of all upper case letters should be avoided for all but short labels such as a title, because they are not as easy to read as upper and lower case letters.

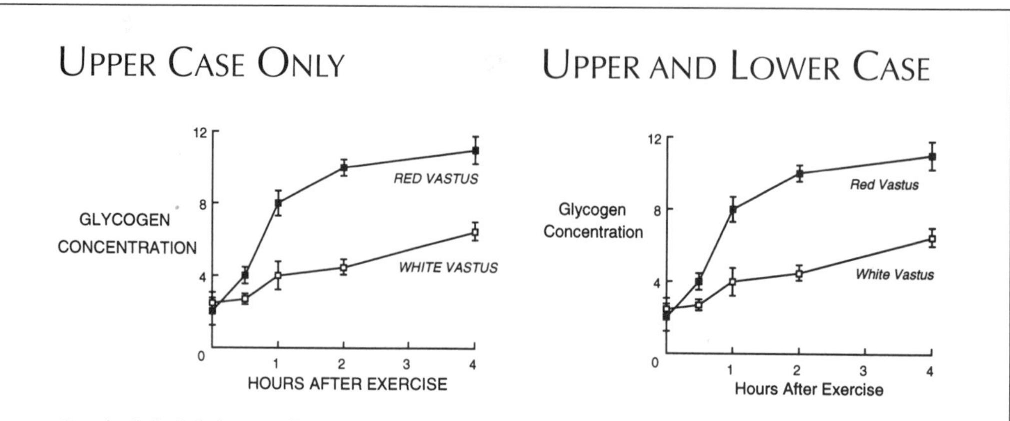

On the left, labels are all upper case. Not only is the same importance given to all labels, but the increased size of the y-axis label enlarges the total graph size. On the right, labels are upper and lower case and varied in size. This sets priorities of importance and is visually more interesting to look at.

Avoid the use of bold or thick labels except for great emphasis. Bold letters tend to dominate the graph, and because they tend to become even thicker in reproduction, they are harder to read.

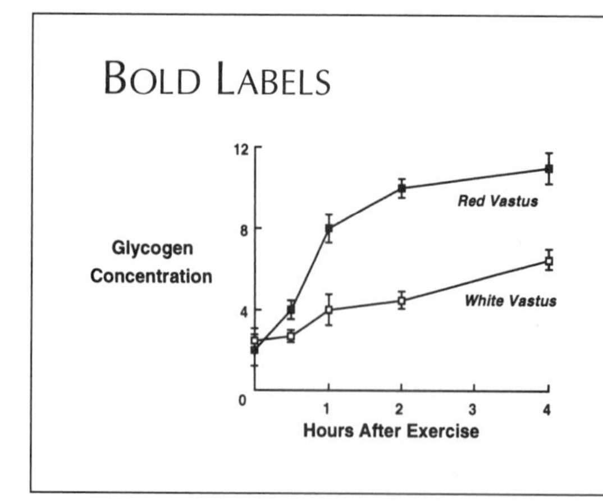

Compare the bold labeling on this graph to the plain labeling of the graphs above. Because labels are now thicker than most of the lines of the graph, they are too emphatic. The letters themselves are blurred and have lost the crispness and openness of plain labels.

LABEL POSITION

Labels may be positioned horizontally or vertically. Although x-axis labels are always horizontal and centered under the axis line, y-axis labels may be either horizontal or vertical. They may be positioned to the side or above the y-axis line. Curve labels may be grouped as a legend or positioned beside each curve.

Additional labels, as for arrows or areas of interest, should be positioned with regard to symmetrical and balanced spatial arrangement. Avoid positioning labels in such a way as to cause confusion or clutter.

Take into consideration whether you will be using your graph for a slide, publication, or poster. Since horizontal labels are easier to read, for slide or poster *all labels* should be horizontal. For publication, the amount of reduction, the page format, and the final shape of the graph will determine whether the y-axis label will be vertical or horizontal.

HORIZONTAL LABELS

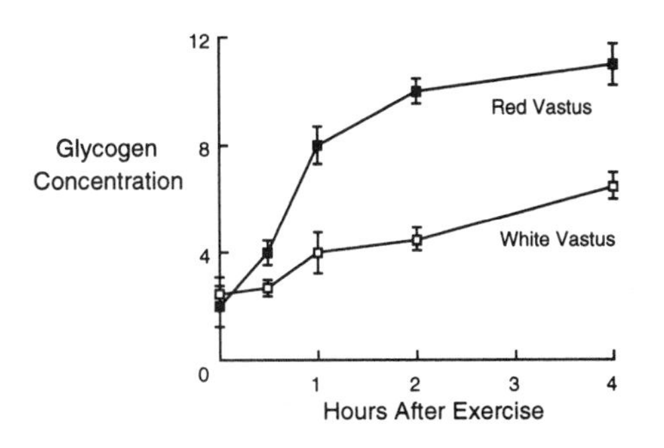

VERTICAL Y-AXIS LABEL

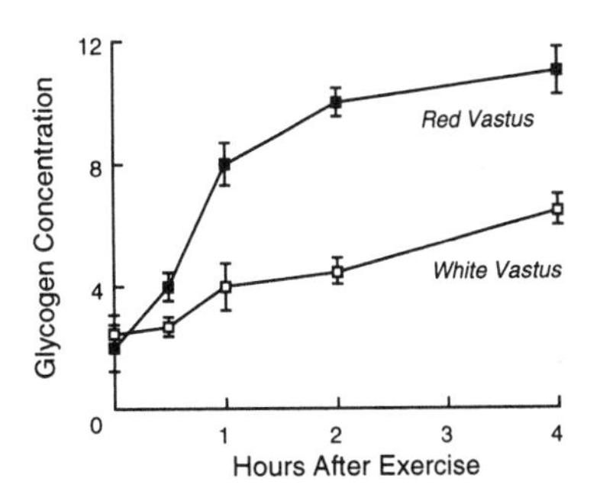

The horizontal labels on the left widen the graph. On the right, the figure is more compact. For a one-column journal, the wider figure with the horizontal labels will stretch across the page and be easily read. For reduction to one column, the squarer shape of the figure on the right will fit the column better. For slides, horizontal labels are always easier to read.

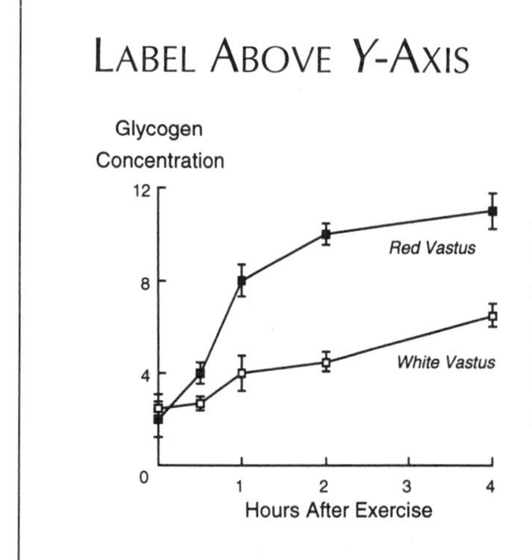

It is not unusual to see the label positioned above the axis line. Long labels may be positioned this way to avoid aligning them vertically. For a slide with a title, however, this would not work well because the axis label and the title would be too close to each other.

Legend Labels

A legend explaining symbols used in a graph is an indirect form of labeling. For identification, the viewer must look from the curve to a legend positioned away from the curve. Identification is much quicker with direct labels positioned next to the curve. However, to avoid repetition of bar labels or symbol labels in composite graphs, a legend can be used.

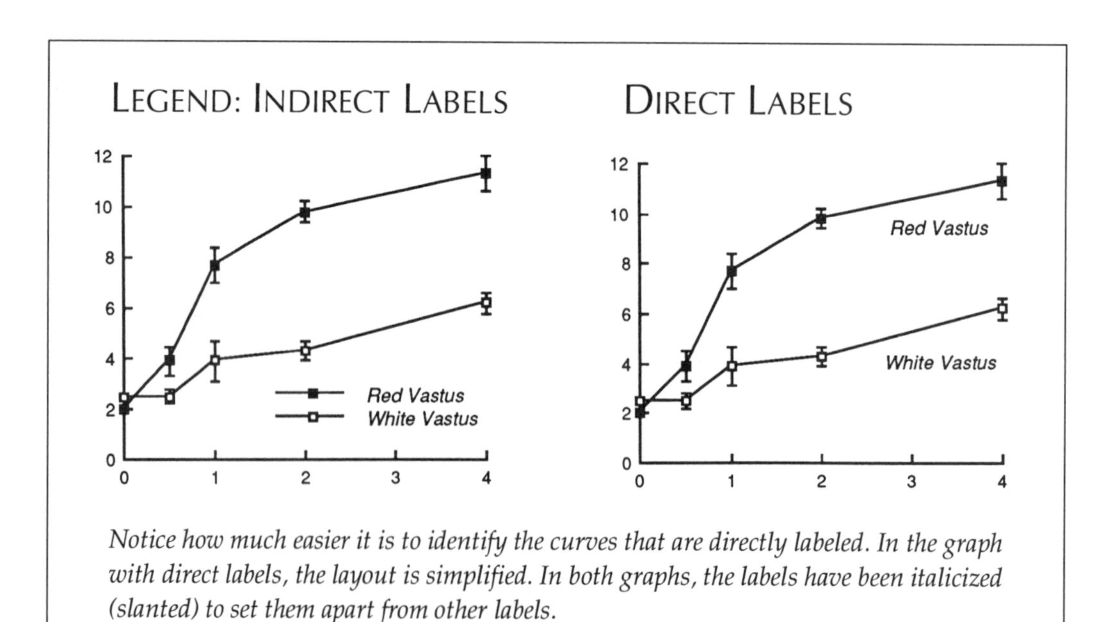

Notice how much easier it is to identify the curves that are directly labeled. In the graph with direct labels, the layout is simplified. In both graphs, the labels have been italicized (slanted) to set them apart from other labels.

Legends are sometimes boxed in; in some computer programs, this is an automatic or default feature.

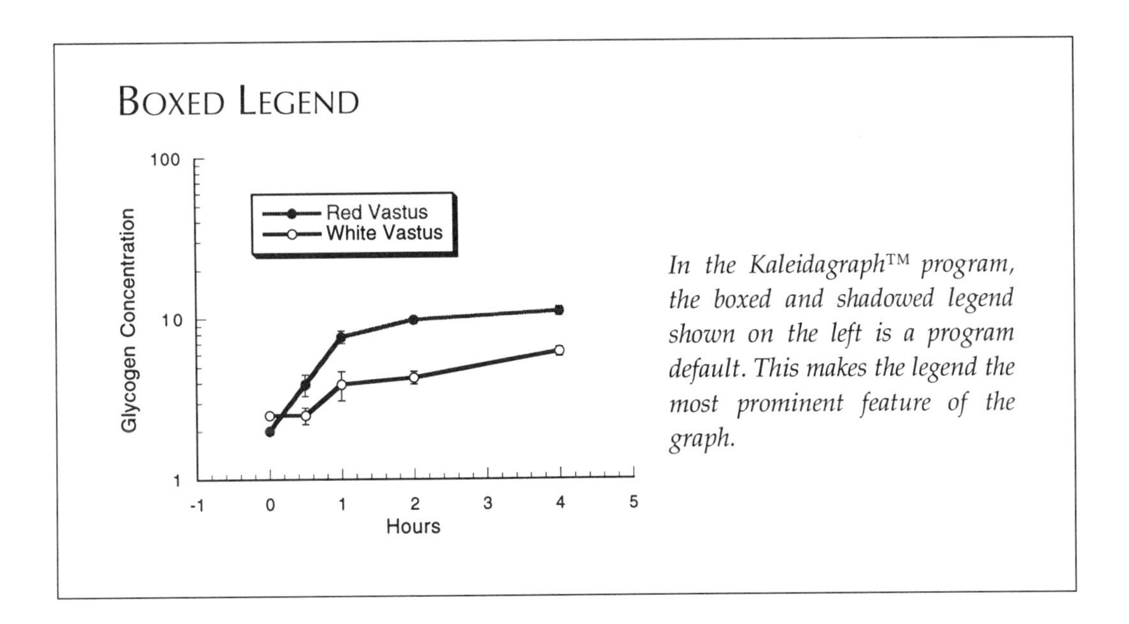

In the Kaleidagraph™ program, the boxed and shadowed legend shown on the left is a program default. This makes the legend the most prominent feature of the graph.

Putting a box around items serves to isolate and emphasize them. Because a legend needs no emphasis, this is not a good idea. Do not add extra lines to a graph unless you have a good, functional reason for it. The simpler and less cluttered your graph is, the better it will communicate.

Labels for Composite Graphs

When putting several graphs together to form one figure (a composite graph), several considerations are important:

- It is not necessary or desirable to repeat labels.
- Be consistent in label size, font, and style.
- Use consistent symbols.
- Design the shapes of the figures to be in conformity with each other and to fit into a format suitable for whatever medium will be used.

COMPOSITE FIGURE

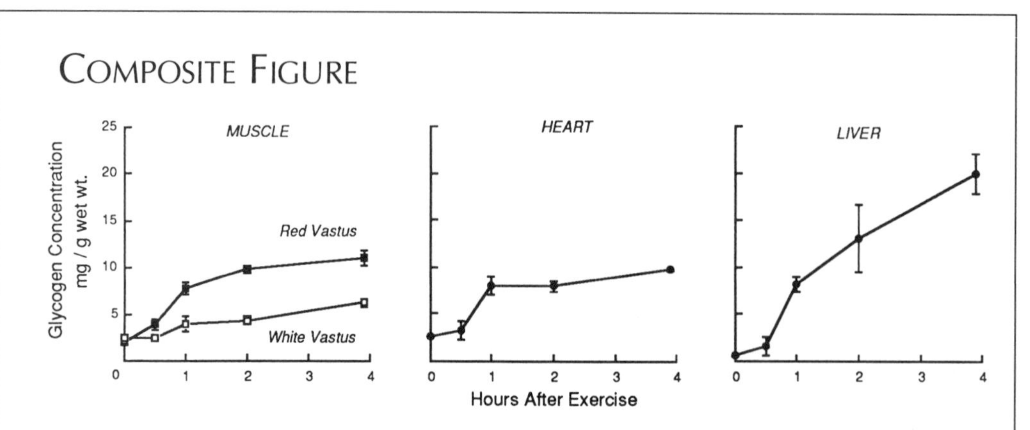

Here, three graphs are shown together as one figure. The x-axes were shortened slightly to accommodate the width of a journal page. Only one y-axis is labeled and numbered, and one label is sufficient for all three x-axes. Each graph has a short title explaining clearly the differences between the three. (These short titles are upper case to distinguish them from the curve labels and italicized to set them apart from the axis labels.)

DESIGN SYMBOLS AND LINES

Design symbols identify variables and distinguish one from the other. The most commonly used are circles, squares, and triangles, either open (unfilled) or closed (filled). Circles and triangles are more distinct from each other than circles and squares, and open and closed symbols show distinctions most clearly. (See the top graph on the opposite page.)

Symbol size should be in proportion to the size of the graph and to graph priorities. In a line graph, the symbols and lines together form a picture of change or flow. Overly large symbols might distract from this picture. In a scattergram with a limited number of points, however, points might be made larger.

Avoid Xs, crosses, or symbols with dots, Xs, or crosses in them. In reduction, they tend to fill up unevenly. Use symbols that are most distinct and use them consistently.

Lines of different thickness and style can reflect priorities by emphasizing or playing down. (See the bottom figure on the opposite page.) Thick lines emphasize; thin lines play down. The solid, dashed, or dotted lines are the most easily distinguished from each other, and of those three, the dotted line is the least emphatic, the solid line most emphatic.

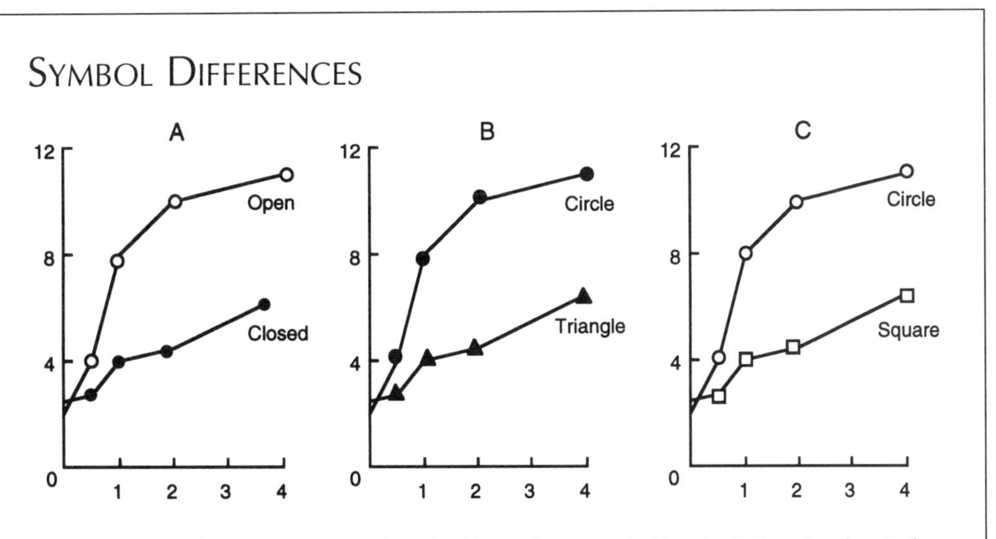

Open and closed design symbols (graph A) are the most distinctly different. The circle and triangle shapes (graph B) are not as distinct as open and closed symbols in graph A. In reduction, the circles and squares (graph C) are the least distinct. Closed symbols are more emphatic than open circles. Triangles appear smaller than circles. In B, the triangle is one size larger than the circle. Squares look larger than circles.

Avoid using different kinds of dashed lines in the same graph because it is difficult to tell them apart. (See the top figure on page 98.) It is not necessary and usually not desirable to vary both lines and symbols. In a series of similar graphs, take care to use the symbols and lines consistently.

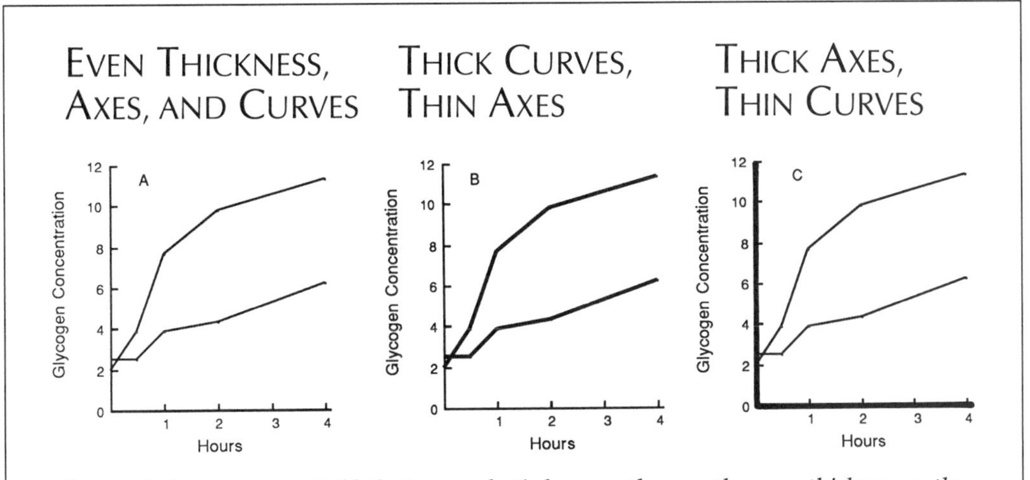

In graph A, curves are visible but unemphatic because they are the same thickness as the axis lines. In graph B, the curves are the first thing the viewer sees. In graph C, the thick axis lines distract from labels and curves and diminish their importance.

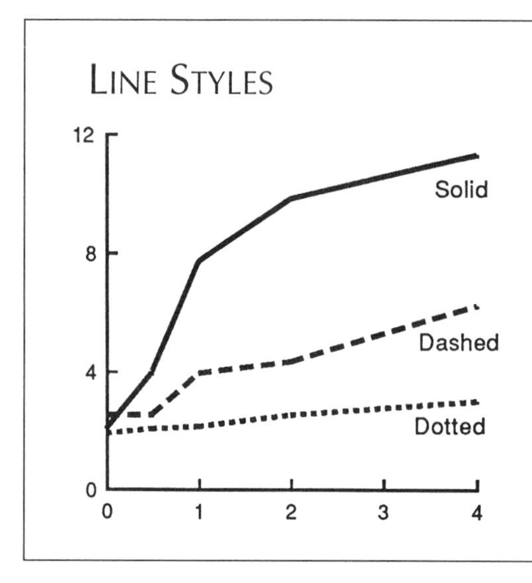

The three line styles with the most contrast are shown here. Use the dotted line for a curve with least significance. If all three curves are of equal significance and do not overlap, it is not necessary to use different line styles.

TEXTURE AND CONTRAST

Texture is a visually interesting way to clarify, emphasize, and distinguish differences. Contrast, as we have already seen from open and closed symbols, accomplishes the same thing.

Stippling (small dots), hatching (lines run vertically, horizontally, or diagonally) and crosshatching are useful devices for identification and encoding. Stippling is also effective in designating areas of interest.

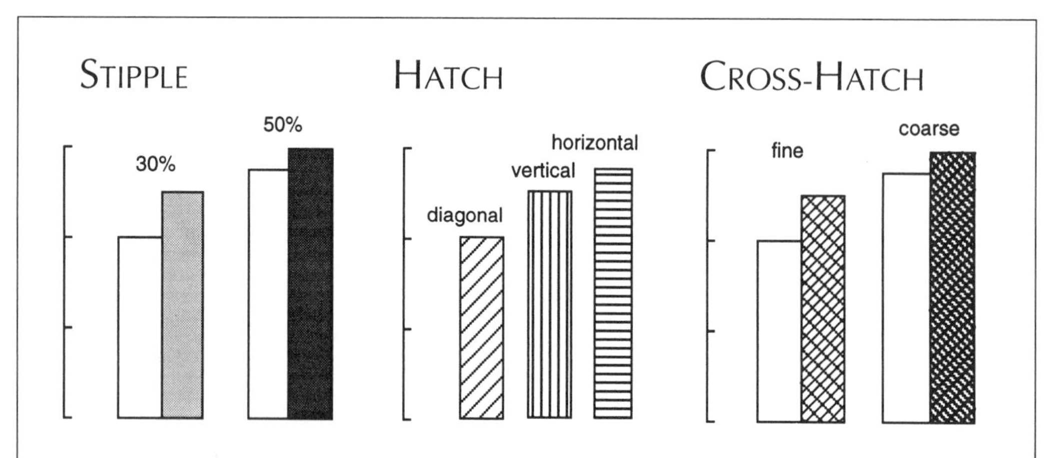

When reduced, stipples become a gray tone. Medium-size dots and middle tones (30–50%) work best in reduction. Very fine stippling tends to drop out or become uneven in reduction. Diagonal hatching is the least ambiguous and most pleasing to the eye. Crosshatching is darker and more emphatic. Use a fine line for crosshatching, because thick-lined crosshatching may fill and become black in reduction.

Graphing software offers a limited variety of suitable patterns and texture, so experiment with the patterns offered to see which have the best contrast and are the most pleasing. Some textures, such as brick and tile patterns, are better for architectural drawings than for graphs.

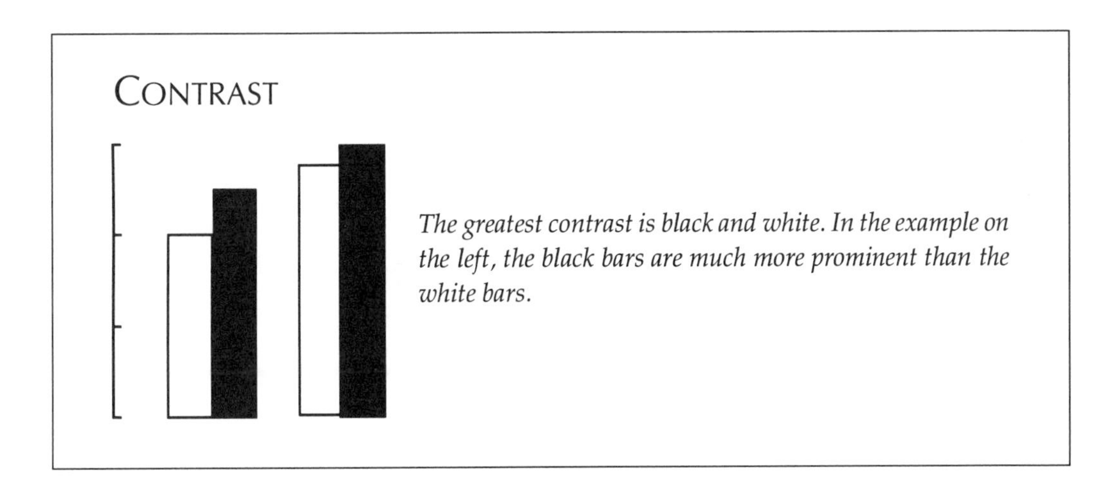

The greatest contrast is black and white. In the example on the left, the black bars are much more prominent than the white bars.

Texture should be used judiciously. Too many kinds of texture in one figure might confuse and distract from the message. Use texture and contrast as an accent or highlight to clarify and simplify as well as to attract the viewer's attention.

ARROWS AND BRACKETS

Arrows are symbols that draw the eyes' attention to an object or event or that denote motion. Arrows are not appropriate for leaders (lines) from labels to the object labeled. Plain, thin lines are best for this.

Arrows effectively indicate occurrence or direction of an event.

Brackets or stippled areas are useful to indicate events that occur over a space of time. (See the figures on the next page.)

The problems discussed in this chapter have to do with clear use of symbolism, proportion, contrast, composition, and texture. These are essential components for communication of sophisticated information. Although they are artists' tools, understanding of and familiarity with their use may be learned from observation and trial.

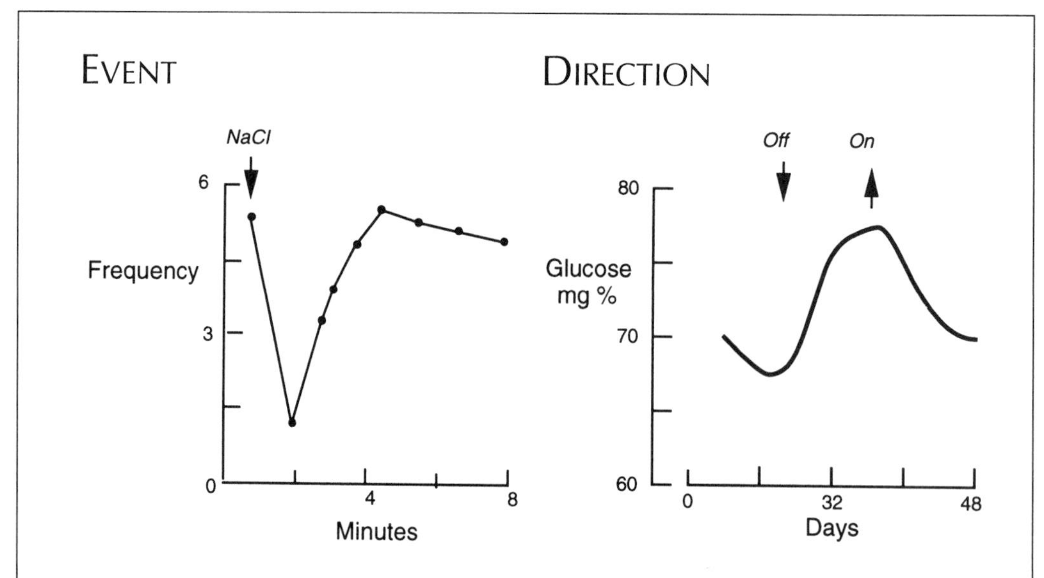

On the left, the arrow shows the time of administration of NaCl. On the right, the up and down directions of the arrows symbolize on and off.

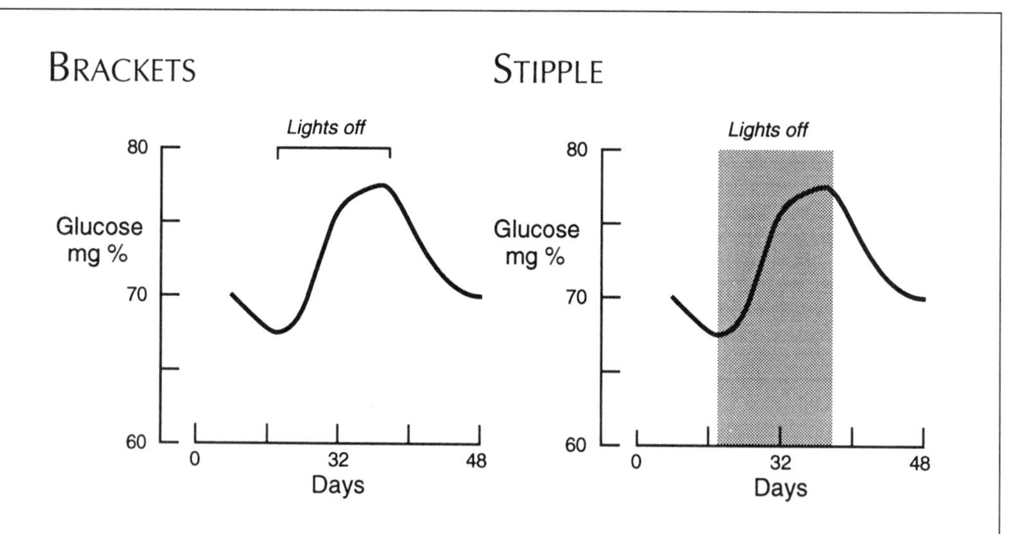

Either brackets or stipple may show an event that occurs over a space of time. The brackets on the left show clearly the duration of the event (lights off). The eye approximates the relationship of the duration to the curve changes. The stippled area on the right crosses the curve, indicating exactly where the event changed the curve.

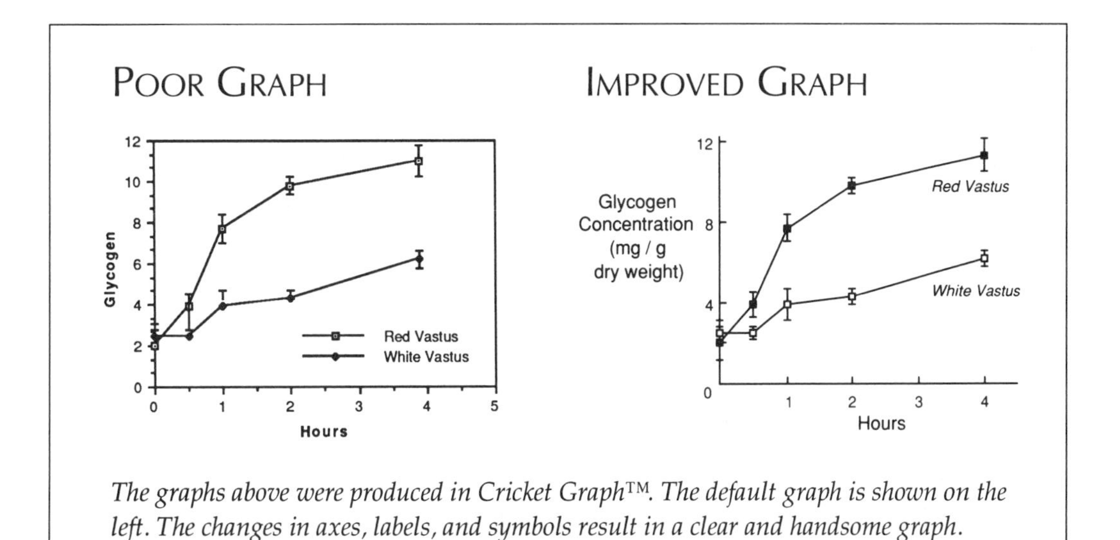

The graphs above were produced in Cricket Graph™. The default graph is shown on the left. The changes in axes, labels, and symbols result in a clear and handsome graph.

FOR GOOD GRAPHS

- Do not put too much information in one graph. Simplify the information.
- Focus is achieved by simplification and graphical emphasis.
- Make the graph appropriate to the medium used: slide, publication, or poster.

Look critically at graphs in journals, slides, and posters. Try to analyze why they are good or bad. Ask yourself how you could improve them.

GRAPHING SOFTWARE

Some of the available graphing software has been illustrated in this chapter. There are good points and shortcomings in all the programs. Familiarity with the software you use may enable you to overcome the limitations of the program. Your own needs are guidelines to the software you use. If you are in the market for a graphing program, below is a list of some of the available programs. Contact the manufacturers for information.

The graphing capabilities of drawing and spreadsheet programs are usually minimal. They are ways for the researcher to visualize numbers, but they do not offer enough graphical options to produce good figures. Transfer data from spreadsheets to graphing programs to achieve a graph that communicates well.

CA Cricket Graph™	Computer Associates International, Inc. 1 Computer Associates Plaza Islandia, NY 11788-7000 Fax: (516) 342-5224; phone: (516) 342-5224
KaleidaGraph™	Synergy Software (PCS Inc.) 2457 Perkiomen Avenue Reading, PA 19606 Fax: (610) 370-0548; phone: (610) 779-0522
Sigma Plot®	Jandel Scientific 2591 Kerner Boulevard San Rafael, CA 94901 Fax: (415) 453-7769; phone: (800) 874-1888
Delta Graph Pro	Delta Point 2 Harris Court, Suite B-1 Monterey, CA 93940 Fax: (408) 648-4021; phone: (408) 648-4000
Harvard Graphics®	Software Publishing Corporation P.O. Box 54983 Santa Clara, CA 95056-0983 Fax: (408) 980-1518; phone: (800) 980-1518

7

The Journal Figure

Just as our clothes fit the weather, figures should fit the medium. A good figure fits into its surroundings. The time and effort spent "dressing" a figure appropriately for its surroundings reflect consideration for the audience and the message.

Although scientists communicate in various media, from a black and white figure for a book to a full-screen, full color movie, the three most commonly used media are the published paper, the lecture, and the poster presentation. Each of these three media have assets and liabilities that must be taken into account.

Throughout the preceding chapters, differences between figures for publication and slide have been pointed out. In this chapter, the special requirements for publication in journals are discussed. Succeeding chapters deal with slides and posters.

The difference between a good published figure and a slide may be only a matter of adding a title or it may be a whole new figure. Whatever it takes, if it gets the message across more quickly and easily, it will be worthwhile.

Publication in journals is expected from scientists and is, if not an everyday occurrence, at least a yearly occurrence. Publication in books, however, is a more momentous and rare occurrence. For this reason, the following pages are specifically directed toward published figures appearing in journals. In most cases, the requirements for books are similar.

A published paper with its illustrations is the most formal and weighty method for disseminating scientific findings. Because the printed page is available for scrutiny by generations of readers, great care should be taken in the writing, and the same care should be taken with the illustrations.

For the reader, this durability and availability are *assets.* A paper may be carried around and read at the reader's convenience. The reader may spend as much or as little time on it as he or she wishes. Details may be included that the reader may study or skip.

The published figure also presents *challenges* that must be met. Chief among these are reduction and accommodation of the figure to limited space and shape. There may also be limitations on the number of figures allowed, and page costs may be a consideration in some journals. Some of or all the cost of color illustrations must be borne by the author and greatly exceed conventional page charges.

The first step to producing an effective journal figure is to know the journal. Instructions to authors in the journal are the first clue.

Journal Instructions

These instructions to authors are not always easy to find, but if they are not in the front or back, look in the table of contents or in the first issue of the volume or write the editorial office. Instructions concerning figures may be a paragraph or a page. They will tell you whether to send originals or photographs; they may inform you of expected reduction and instruct you as to labeling style and size and the kinds of symbols to be used. Whether instructions are sparse or detailed, be sure to read them. Also, look through the journal, noticing such things as number of columns, style of labeling, figure reduction, and figure layout.

Reduction

The final reduction of a figure is not always predictable. It depends on journal size, format, and the editor. However, if you design for maximum reduction within the journal format, your figure should be legible. Maximum reduction usually means reduction to one-column width of a two- or three-column page (2⅜ inches to 3½ inches; 6 cm to 9 cm). In a one-column journal, however, final reduction will depend on the shape of the graph.

The journal may instruct you to design the figure to fit the page economically. Because wasted space costs the journal money, the object of the editor and layout artist is to reduce and position the figure within the surrounding text in a way that is compact not only on the page but in the context of the whole article. This may result in a final reduction to less than one-column width.

In a journal with *two columns,* a square figure will usually suffer less reduction than a long narrow figure.

ONE-COLUMN WIDTH

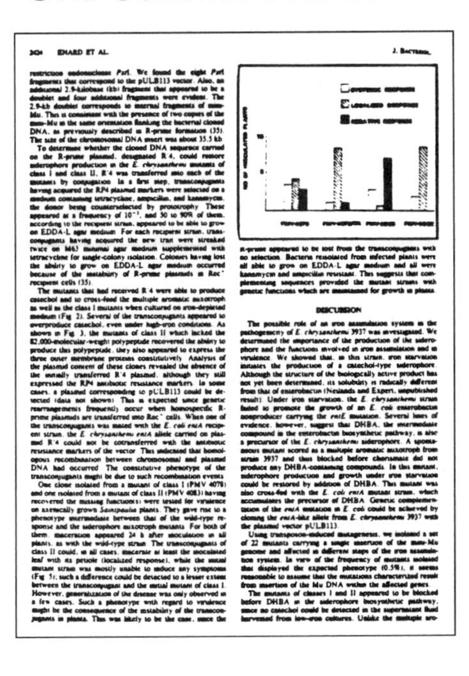

This figure is close enough to being square to occupy all the one-column width. The letter size is about the same size as the text and is easily read. Bars are easily differentiated.

On the following page is a figure that has unused space around it.

LESS THAN ONE-COLUMN WIDTH

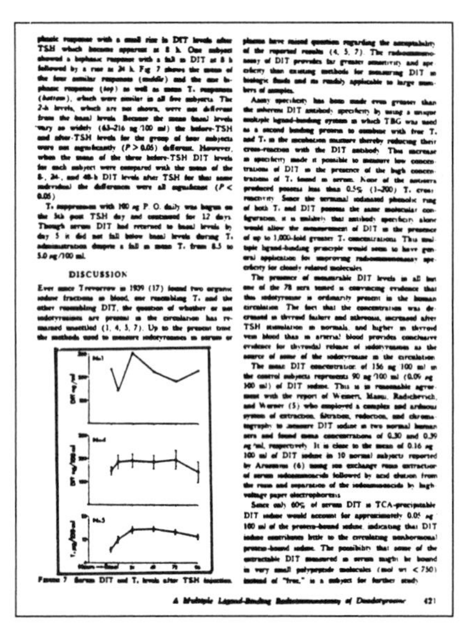

ONE-COLUMN WIDTH

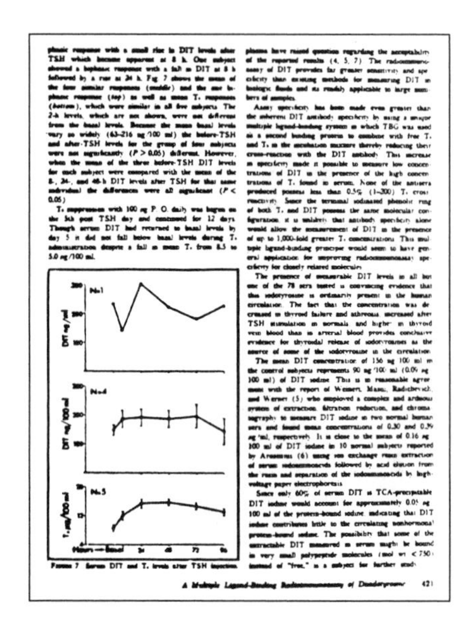

On the left, the composite figure has been reduced to less than one-column width. Notice the unused space on the sides of the figure. This reduction is reasonable because labels are legible and points and lines are easy to see. On the right, the figure has been enlarged to fill the entire column width. Notice the corresponding height of the figure, which now occupies more than half of the column. The left-hand figure above shows how it actually appeared in the journal. In the interest of saving space, the journal reduced the figure to take up less than one-column width. On such a long figure, there is room to arrange the y-axis labels horizontally for easier reading and pleasing composition.

Design your figures for the appropriate reduction.

To predict figure reduction, measure the figure dimensions. Make a box of these dimensions and draw a diagonal line from the lower left corner to the upper right corner of the box. Measure the journal column width and fit that measurement from the left-hand axis to the diagonal line. Draw a vertical line from that point on the diagonal line to the lower axis. You can see how much vertical space your figure will occupy in the column.

Avoid exaggeratedly long or wide shapes. When plotting graphs, make the axis lengths as even as possible.

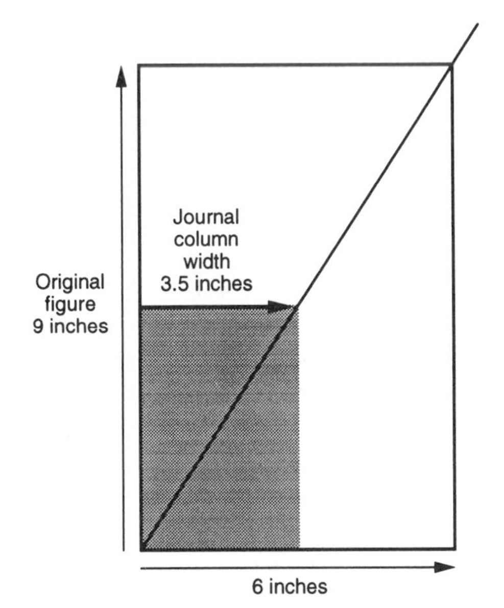

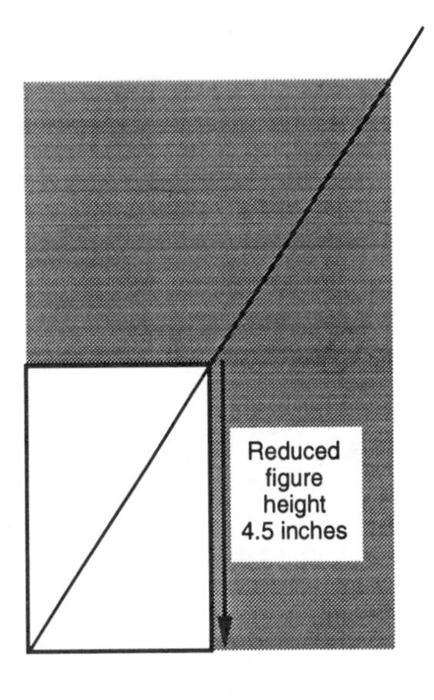

The original figure on the left measures 6 × 9 inches including the labels. The shaded rectangle indicates the width of one column in the journal for which it is intended. On the right, the open box indicates the height of the reduced figure, derived by drawing a vertical line from the measured column width point on the diagonal line to the horizontal baseline. If the reduced figure height occupies half or close to half of the entire column height, as this one does at 4.5 inches, it is safe to assume that it will be reduced still further to conserve space.

FORMAT

Design the figure for the journal. The journal format (page size and shape and number of columns) plays a role in determining reduction. For this reason and for a figure to be maximally effective, consideration of journal format is essential in planning the shape of the figure.

One-Column Journal

The one-column journal is perhaps the most difficult to design for. The most suitable figures are those that are wider than they are long.

PAGE EXAMPLE

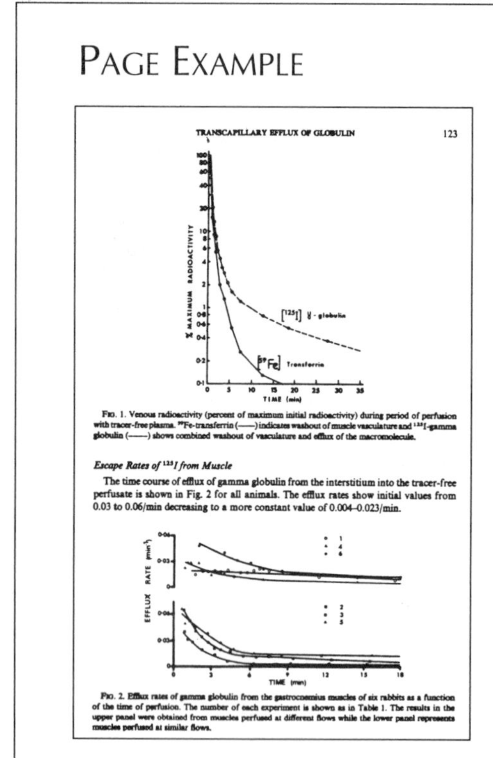

TRANSCAPILLARY EFFLUX OF GLOBULIN 123

FIG. 1. Venous radioactivity (percent of maximum initial radioactivity) during period of perfusion with tracer-free plasma. ^{59}Fe-transferrin (——) indicates washout of muscle vasculature and ^{125}I-gamma globulin (——) shows combined washout of vasculature and efflux of the macromolecule.

Escape Rates of ^{125}I from Muscle

The time course of efflux of gamma globulin from the interstitium into the tracer-free perfusate is shown in Fig. 2 for all animals. The efflux rates show initial values from 0.03 to 0.06/min decreasing to a more constant value of 0.004–0.023/min.

FIG. 2. Efflux rates of gamma globulin from the gastrocnemius muscles of six rabbits as a function of the time of perfusion. The number of each experiment is shown as in Table 1. The results in the upper panel were obtained from muscles perfused at different flows while the lower panel represents muscles perfused at similar flows.

There is more wasted space around the tall top figure than the wide bottom figure. Both figures have the same reduction, but the wide bottom figure wastes less space and has more pleasing proportions on the page. The top figure would benefit from an expanded x-axis. Both figures have room for horizontal labels on the y-axis.

An unsatisfactory alternative to wasting space with a tall figure in a one-column journal is to enlarge it to fit the column or page width. This can be seen in the top figure on the facing page.

An excellent way to save space in a one-column journal is to design side-by-side composite figures.

Figure Too Large

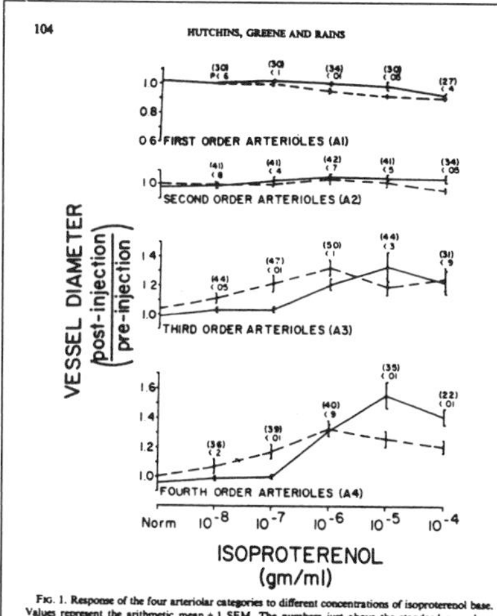

FIG. 1. Response of the four arteriolar categories to different concentrations of isoproterenol base. Values represent the arithmetic mean ± 1 SEM. The numbers just above the standard error bars represent p values computed from a two-tailed t test. The numbers in parentheses just above the p values are the number of observations. The dashed lines represent the SHR and the solid lines, NCR.

hypertensive rats. He also noted that adenylate cyclase activity was normal but less responsive to stimulation in the hypertensive group. Spector *et al.* (1969) also report that aortic strips from SHR demonstrate a greater relaxation to 10^{-7} g/ml of isoproterenol, a finding which was confirmed in the smaller arterioles of this study.

The smaller dilatory capacity in response to isoproterenol may be due to the SHR arterioles being partially dilated. We have previously reported (Hutchins and Darnell, 1974) that the small arterioles of the SHR have larger diameters than their NCR counterparts but are fewer in number. The effect of a prior existing, partial dilation

Although this figure fits the column width, it takes up most of the page. The label size is much larger than the surrounding text, giving it an exaggerated prominence. At half this size, it could still be easily read. The space between the two bottom graphs could be reduced, and the y-axis label could be arranged horizontally on four lines to fit the space better.

One-Column Composite

12 NAKAMURA AND WAYLAND

the fourth. Three spotty sources of diffusion are noted at 5 sec, with the tissue concentration at grades 1 and 2. Although the fluorescent patterns diffusing from two of the sources became confluent by 10 sec, corresponding to Figs. 5 and 7, the spotty sources are still evident from the high tissue concentration near the capillary. A rather irregular spreading pattern is noted with increase in time, and lateral spreading is apparent.

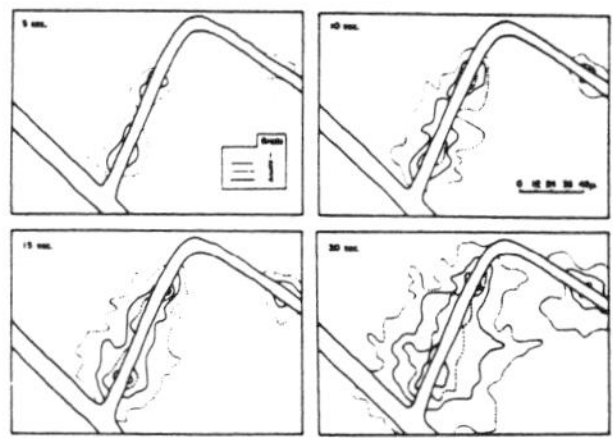

FIG. 8. Contour maps of fluorescent emission of BSA–FITC in the microbed shown in Figs. 5 and 7 as a function of time from first appearance of fluorescence in the field. Note the spreading of the fluorescence in the tissue from the three "hot spots" on the loops.

Apparent Diffusion Coefficients in the Tissue

Although we were were unable to make a quantitative measurement of the concentration of the tagged molecules within the blood vessels, it was possible to calculate an effective diffusion coefficient within the tissue. Because of the unknown thickness of the actual space in the tissue through which the molecular species under study was moving, it is impossible to measure the true concentration in the tissue. If, however, we assume that the thickness of the interstitial space is constant, then the concentration at any point can be normalized against that at a preselected origin near the vessel. As long as the concentration contours are essentially parallel to the vessel, Eq. (5) can be used for the calculation of an apparent diffusion coefficient. The exact location of the origin is not important as long as the history of the concentration at this locus is known.

Figure 9 shows a series of concentration profiles for FITC-BSA as a function of distance from a venous capillary for various times. The ordinate represents an arbitrary scale, but is the same for all curves. It is seen that $C(0, t)$, that is the concentration just outside the vessel wall, increases with time, first rapidly, and then more slowly. The increase of concentration with time at the origin was essentially linear for the first 10 sec, permitting the use of Eqs. (6) or (8) for calculating a diffusion coefficient.

The normalized concentration profile for FITC-BSA just outside a venous capillary for the 10-sec data of Fig. 9 is shown in Fig. 10, where the ordinate is the ratio of the

There is little wasted space in this figure because the overall shape is wider than it is tall. A wide figure has better proportions for one-column journal format and wastes less space.

It is not always possible to design wide figures or to combine figures side by side. But when planning figures, keep in mind journal format. Make figures conform as closely as possible to the space.

Two-Column Journal

There are more options for figure position and shape in a two-column journal. The figure may be reduced to one column or it may stretch across the page. In rare instances, it may occupy an entire page.

ONE- AND TWO-COLUMN WIDTHS

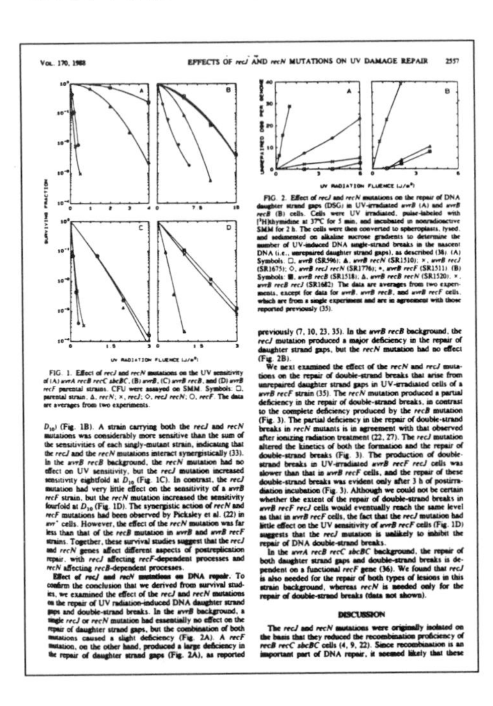

VOL. 170, 1988 EFFECTS OF *recJ* AND *recN* MUTATIONS ON UV DAMAGE REPAIR 2557

FIG. 1. Effect of *recJ* and *recN* mutations on the UV sensitivity of (A) *uvrA recB recC sbcBC*, (B) *uvrB*, (C) *uvrB recB*, and (D) *uvrB recF* parental strains. CFU were assayed on SMM. Symbols: □, parental strain; Δ, *recN*; ×, *recJ*; ◇, *recJ recN*; ○, *recF*. The data are averages from two experiments.

D_{10}) (Fig. 1B). A strain carrying both the *recJ* and *recN* mutations was considerably more sensitive than the sum of the sensitivities of each singly-mutant strain, indicating that the *recJ* and the *recN* mutations interact synergistically (33). In the *uvrB recB* background, the *recN* mutation had no effect on UV sensitivity, but the *recJ* mutation increased sensitivity eightfold at D_{10} (Fig. 1C). In contrast, the *recJ* mutation had very little effect on the sensitivity of a *uvrB recF* strain, but the *recN* mutation increased the sensitivity fourfold at D_{10} (Fig. 1D). The synergistic action of *recN* and *recF* mutations had been observed by Picksley et al. (22) in *uvr*⁺ cells. However, the effect of the *recN* mutation was far less than that of the *recB* mutation in *uvrB* and *uvrB recF* strains. Together, these survival studies suggest that the *recJ* and *recN* genes affect different aspects of postreplication repair, with *recJ* affecting *recF*-dependent processes and *recN* affecting *recB*-dependent processes.

Effect of *recJ* and *recN* mutations on DNA repair. To confirm the conclusion that we derived from survival studies, we examined the effect of the *recJ* and *recN* mutations on the repair of UV radiation-induced DNA daughter strand gaps and double-strand breaks. In the *uvrB* background, a single *recJ* or *recN* mutation had essentially no effect on the repair of daughter strand gaps, but the combination of both mutations caused a slight deficiency (Fig. 2A). A *recF* mutation, on the other hand, produced a large deficiency in the repair of daughter strand gaps (Fig. 2A), as reported

FIG. 2. Effect of *recJ* and *recN* mutations on the repair of DNA daughter strand gaps (DSG) in UV-irradiated *uvrB* (A) and *uvrB recB* (B) cells. Cells were UV irradiated, pulse-labeled with [³H]thymidine at 37°C for 5 min, and incubated in nonradioactive SMM for 2 h. The cells were then converted to spheroplasts, lysed, and sedimented on alkaline sucrose gradients to determine the number of UV-induced DNA single-strand breaks in the nascent DNA (i.e., unrepaired daughter strand gaps), as described (38). (A) Symbols: □, *uvrB* (SR596); ▲, *uvrB recN* (SR1510); ×, *uvrB recJ* (SR1675); ◇, *uvrB recJ recN* (SR1776); •, *uvrB recF* (SR1511). (B) Symbols: ■, *uvrB recB* (SR1518); Δ, *uvrB recB recN* (SR1520); ×, *uvrB recB recJ* (SR1682). The data are averages from two experiments, except for data for *uvrB*, *uvrB recB*, and *uvrB recF* cells, which are from a single experiment and are in agreement with those reported previously (35).

previously (7, 10, 23, 35). In the *uvrB recB* background, the *recJ* mutation produced a major deficiency in the repair of daughter strand gaps, but the *recN* mutation had no effect (Fig. 2B).

We next examined the effect of the *recN* and *recJ* mutations on the repair of double-strand breaks that arise from unrepaired daughter strand gaps in UV-irradiated cells of a *uvrB recF* strain (35). The *recN* mutation produced a partial deficiency in the repair of double-strand breaks, in contrast to the complete deficiency produced by the *recB* mutation (Fig. 3). The partial deficiency in the repair of double-strand breaks in *recN* mutants is in agreement with that observed after ionizing radiation treatment (22, 27). The *recJ* mutation altered the kinetics of both the formation and the repair of double-strand breaks (Fig. 3). The production of double-strand breaks in UV-irradiated *uvrB recF recJ* cells was slower than that in *uvrB recF* cells, and the repair of these double-strand breaks was evident only after 3 h of postirradiation incubation (Fig. 3). Although we could not be certain whether the extent of the repair of double-strand breaks in *uvrB recF recJ* cells would eventually reach the same level as that in *uvrB recF* cells, the fact that the *recJ* mutation had little effect on the UV sensitivity of *uvrB recF* cells (Fig. 1D) suggests that the *recJ* mutation is unlikely to inhibit the repair of DNA double-strand breaks.

In the *uvrA recB recC sbcBC* background, the repair of both daughter strand gaps and double-strand breaks is dependent on a functional *recF* gene (36). We found that *recJ* is also needed for the repair of both types of lesions in this strain background, whereas *recN* is needed only for the repair of double-strand breaks (data not shown).

DISCUSSION

The *recJ* and *recN* mutations were originally isolated on the basis that they reduced the recombination proficiency of *recB recC sbcBC* cells (4, 9, 22). Since recombination is an important part of DNA repair, it seemed likely that these

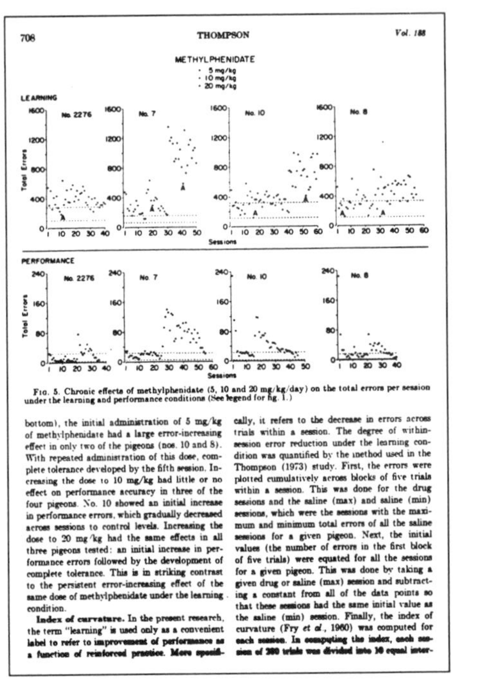

708 THOMPSON Vol. 188

FIG. 5. Chronic effects of methylphenidate (5, 10 and 20 mg/kg/day) on the total errors per session under the learning and performance conditions (See legend for Fig. 1.)

bottom), the initial administration of 5 mg/kg of methylphenidate had a large error-increasing effect in only two of the pigeons (nos. 10 and 8). With repeated administration of this dose, complete tolerance developed by the fifth session. Increasing the dose to 10 mg/kg had little or no effect on performance accuracy in three of the four pigeons. No. 10 showed an initial increase in performance errors, which gradually decreased across sessions to control levels. Increasing the dose to 20 mg/kg had the same effects in all three pigeons tested: an initial increase in performance errors followed by the development of complete tolerance. This is in striking contrast to the persistent error-increasing effect of the same dose of methylphenidate under the learning condition.

Index of curvature. In the present research, the term "learning" is used only as a convenient label to refer to improvement of performance as a function of reinforced practice. More specifically, it refers to the decrease in errors across trials within a session. The degree of within-session error reduction under the learning condition was quantified by the method used in the Thompson (1973) study. First, the errors were plotted cumulatively across blocks of five trials within a session. This was done for the drug sessions and the saline (max) and saline (min) sessions, which were the sessions with the maximum and minimum total errors of all the saline sessions for a given pigeon. Next, the initial values (the number of errors in the first block of five trials) were equated for all the sessions for a given pigeon. This was done by taking a given drug or saline (max) session and subtracting a constant from all of the data points so that these sessions had the same initial value as the saline (min) session. Finally, the index of curvature (Fry *et al.*, 1960) was computed for each session. In computing the index, each session of 280 trials was divided into 16 equal inter-

The composite figures on the left fill one-column widths. The figure on the right stretches across the page, occupying two columns and the center space between the columns.

Three-Column Journal

Although two-column journals are the most common, three-column journals are becoming increasingly frequent. There are even more options for figure shape and position with three columns. In addition to occupying one, two, or three columns, a figure may spread over parts of several columns.

THREE-COLUMN PAGE

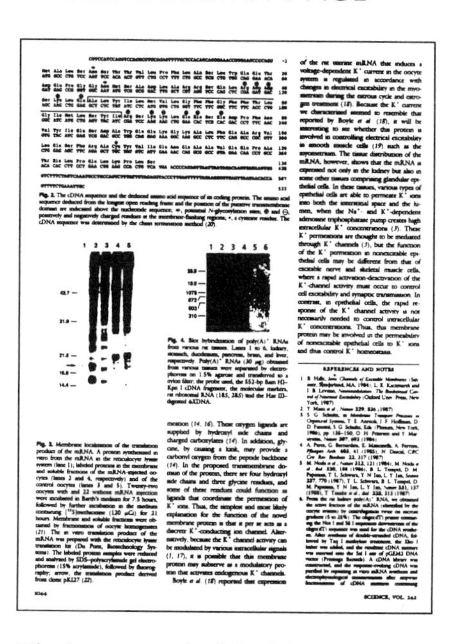

The layout on the left shows a top figure occupying two columns and, below it, two one-column figures. On the right, the top figure occupies one and a half columns with the caption wrapping around it. Below it, the gel occupies half of a column surrounded by its caption.

There are many variations in figure layout for different formats. Although the researcher does not have a great deal of control over the layout, it is important to be aware of the possibilities, keeping in mind the fact that every journal aims for a compact and space-saving layout within whatever format is used.

LABELS

Figure *titles* are not used in journals. The information in a title, along with a more detailed figure explanation, is contained in the caption beneath the figure. The *caption* itself is an integral part of the final published figure on the page. Although occasionally a caption will be placed on the opposite page from the figure, as with a full-page photographic plate, it is generally positioned close to the figure. It may even wrap itself around the figure, as can be seen in the previous example.

Label Position

The position of the label on the x-axis is always horizontal, although for the y-axis it may be vertical or horizontal. Horizontal labels are always easier to read, but if a y-axis label is long and cannot easily be broken into several horizontal lines, it should be positioned vertically.

HORIZONTAL OR VERTICAL LABELS?

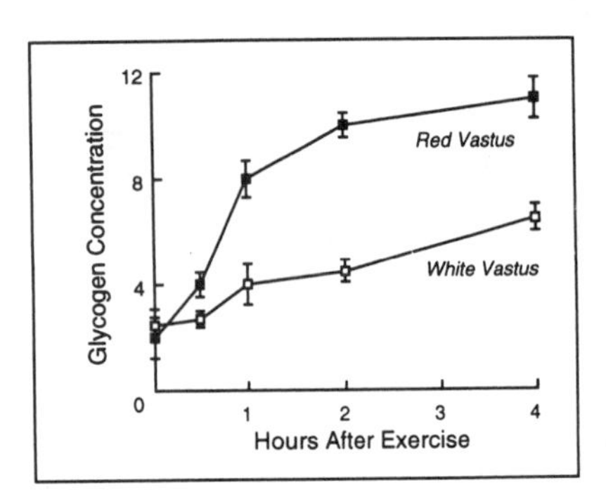

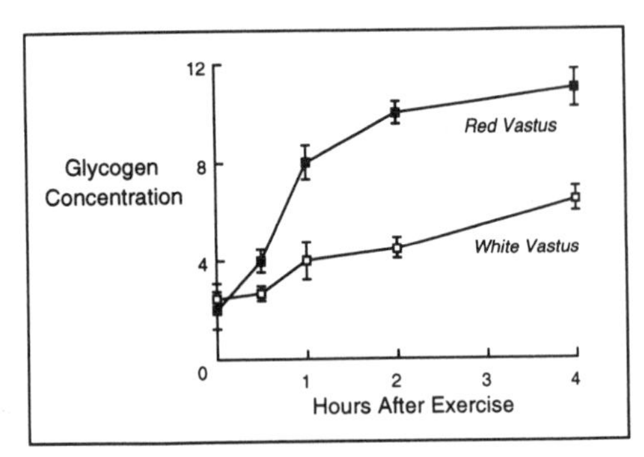

On the left, the y-axis label is vertically aligned. This maintains the almost square shape of the figure and is suitable for reduction to one column of a two-column journal. For a one-column journal, however, the horizontal alignment shown on the right is preferable because it creates a wider figure, which fills a one-column space better.

LABEL ABOVE THE Y-AXIS

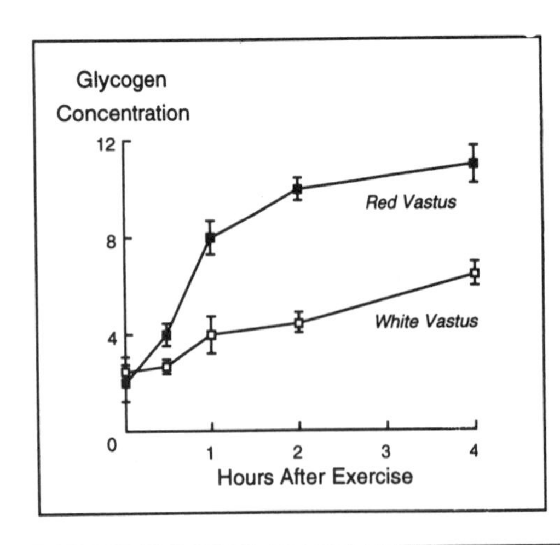

This y-axis label is divided into two lines. The square shape of the figure makes it suitable for reduction to one column while retaining the easy-to-read horizontal labels. If a label is so long that it occupies three lines above the y-axis, it would be better to position it vertically. As well as looking unwieldy, three lines of labeling will elongate the figure.

The lower figure on the facing page shows another option, positioning the label above the *y*-axis. This is a way to maintain the ease of horizontal reading while still using space economically.

Label Style and Size

Label size in the reduced figure should not be smaller than the print on the page. Nor should it be much larger. Journal text is 8- to 10-point size, so ideally the reduced label size should not exceed 12 point. This means that if the original will be reduced by half, the labels on the original should be 24 point. Size of numbers and subsidiary labels may be smaller (18 point). For photomicrographs, labels may be larger (14 to 20 point) because of the difficulty of seeing smaller labels on such figures.

APPROPRIATE LABEL SIZE

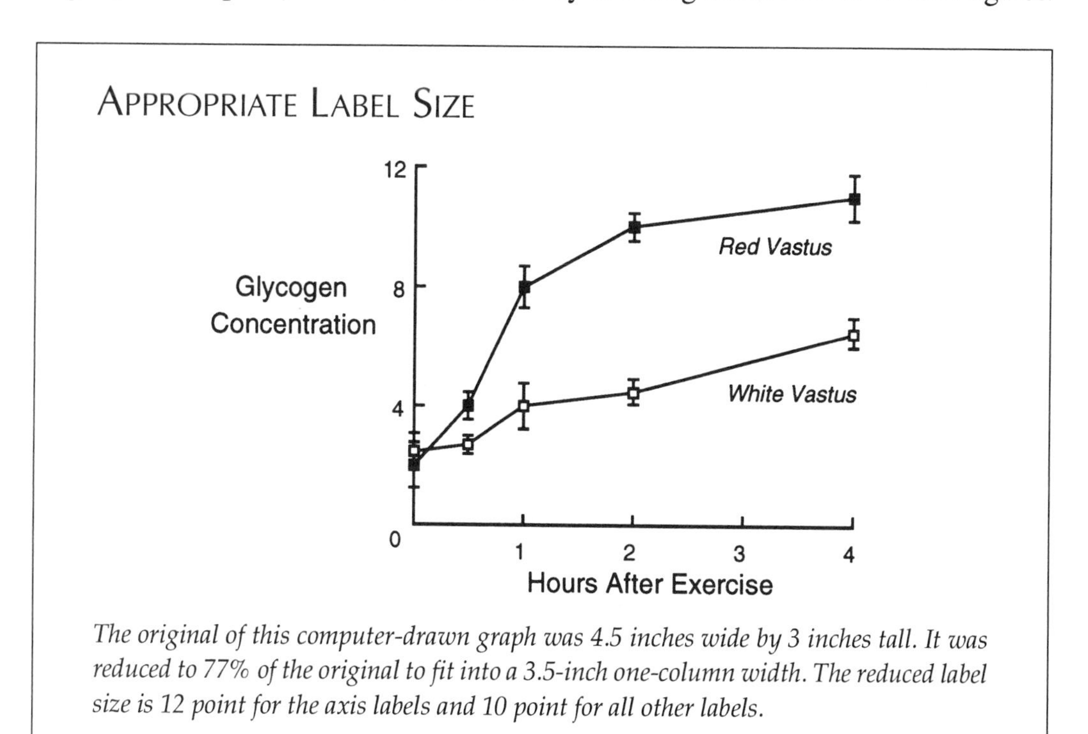

The original of this computer-drawn graph was 4.5 inches wide by 3 inches tall. It was reduced to 77% of the original to fit into a 3.5-inch one-column width. The reduced label size is 12 point for the axis labels and 10 point for all other labels.

For the preceding graphs, label style is plain for the numbers and axis labels, italics for the curve labels. Helvetica font is used throughout. To check for legibility of labels, reduce the figure on a photocopying machine to the required journal dimensions.

Some journals require that legends or explanatory labels be described in the caption. Others ask that there be no numbers or letters demarcating composite graphs one from the other. There are even some journals that ask that there be no labels. Check the journal instructions before labeling the figures.

CONSISTENCY

Consistency of size, shape, nomenclature, and symbols is extremely important to the smooth flow and easy understanding of information. In planning your figures, think of them in terms of the whole paper and design figures that are uniform in size.

If figures are made at different times or are from various sources, consider adapting them to conform to a consistent style.

FINAL PREPARATION

After the figures are completed and before any required photography, check them carefully for errors. Ask colleagues to check them for errors. Changes made after photography are costly and time-consuming.

Examine the final figure carefully for consistently black lines and clean white background. Reduce the figure on a copy machine to check label legibility and to make sure that lines do not drop out and that stippled areas do not fade.

If glossy photographs are submitted, check them for good contrast and sharpness of line. In the case of continuous-tone illustrations, check the tonal range and faithfulness to the original.

If you want figures to appear near each other or with specific text, indicate this clearly on a cover sheet.

Some journals allow computer-printed figures instead of glossy prints. Ask the journal what its policy is about computer-drawn figures. It may be possible to send the disks of your figures for the journal to print from. The fewer processes an original figure goes through to reach the final print, the less possibility there will be for loss in quality.

When you mail originals, put a cover sheet over them or place them in a separate envelope to keep them clean and intact. If you mail photographic prints, avoid indentations made by paper clips or ball-point pen labels on the back. When mailing, use heavy cardboard to protect figures from bending.

All figures must be numbered consecutively to conform to the text and caption. Check to make sure numbering is correct. Number lightly on the cover sheet or on the back of the figure, or number on gummed labels applied to the figure back.

THE PRINTING PROCESS

When the paper is typeset, blank space is left on the page into which the line figures are inserted. For each line figure, a high-quality positive print (called a photomechanical transfer print, or PMT, sometimes called a "stat") is made in the final size, and this PMT is pasted into the opening. The PMT made from high-contrast or line art records only black and white. A negative of this whole page with type and PMT in place is then made and used to make the final printing plate.

Continuous-tone materials (photographs) are handled differently, because they must be converted into halftones (screened). When the page is typeset, an opening (window) is left in the size in which the final halftone is to be printed. A negative is made of this whole page, and the halftone negative that has been made in a separate process is fitted into the window negative of the text page. Once all page negatives are completed, their images are transferred to a metal printing plate. The plates are attached to the printing press, inked, and run in the offset process.

Color

One reason why color illustrations are so expensive is that the process for producing color figures is complicated and demands special care and high technology. The cost of a color plate in a journal must be borne by the author.

A separate printing plate must be made for each color. If *spot color* (one color) is used, as for specific columns in a bar graph, the artist must prepare a separate overlay for the color area. This overlay must be in exact registration with the black and white original, and the desired color must be indicated. Separate negatives are made of the original and the overlay, and separate plates are made from these negatives.

For *full color* reproduction, as of a photograph or full color drawing, the original must be scanned either mechanically or digitally using filters to separate the colors. The scanning process is analogous to the halftone process in that tiny dots are created to generate the final printed image. All the colors of the original are created by mixing dots in various proportions of three printing colors plus black. A negative is made for each color and black; all four of these negatives must be in precise register. Often color corrections must be made, modifying the negative. Four metal printing plates are then made from the four negatives, which are put on the printing press in register. When the paper moves through the press, an image from each color is impressed on the paper.

Proofs

Journals usually send proofs to the author. *Galley proofs* are made before final page layout has been done and show text in long blocks of type with reduced figures grouped separately. However, to save time, most journals now skip galley proofs and go directly to *page proofs*. The author sees photocopies of the final typeset pages with final size line art in place and windows (blank spaces) where halftones are to appear.

Halftones, reduced as they will appear in print, are supplied on separate sheets at the end of the manuscript. The quality of proof reproduction is not as good as in the final print appearing in the journal.

At the page proof stage, it is tedious and expensive to change the position or size of a figure. Special requirements for figures should be indicated on the original manuscript.

> Any changes that you make that correct your own errors will be at your expense. The journal pays for errors it commits. Proof the figures and text carefully for errors.

8 SLIDES

A slide is usually seen for less than 30 seconds, so its impact has to be immediate. For this reason, figures for slides must be especially simple and succinct. A good slide makes no more than three points, and these points augment, emphasize, and explain the speaker's words.

Sometimes speakers try to make one slide do the work of many. The result is visual confusion. For complicated subject matter, it is far better to use two or three simple figures than one complex and usually cluttered and unclear figure. Your audience may fall asleep while you are spending minutes trying to clear the confusion.

COMPLICATED SLIDE

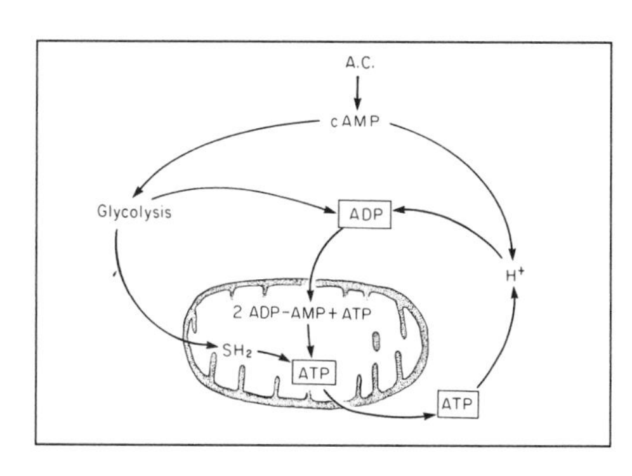

This slide will take considerable explanation and is not clear on first view. Although all the information shown here is necessary to describe the cyclical process, it could be presented more gradually and simply.

SLIDE 1

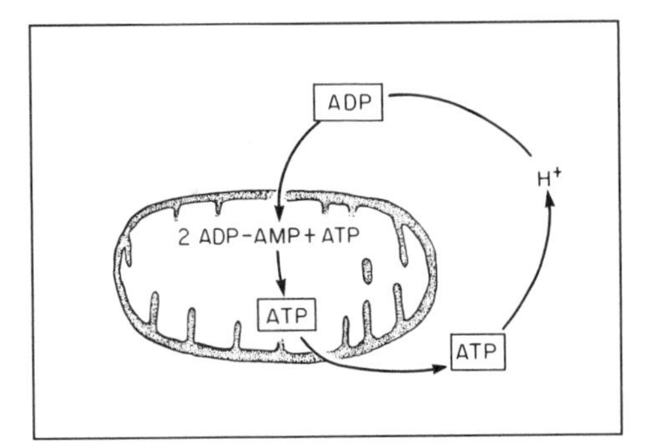

This slide shows one ATP-ADP cycle. The information here can be grasped quickly, so that the slide need not be left on for long. This and the following slide figures are shown in an appropriate slide format: a horizontally aligned rectangle.

In addition, use of a series of slides that build on each other helps to create a dynamic and suspenseful flow of information. This can be a most effective teaching tool.

Starting simply and gradually adding information is a much more dynamic and interesting way to communicate. Notice that every addition to the three-slide series on this page and the following page increases the overall size of the figure, resulting in increased reduction.

An overwhelming amount of information when presented at one time can be easily assimilated when presented progressively.

SLIDE 2

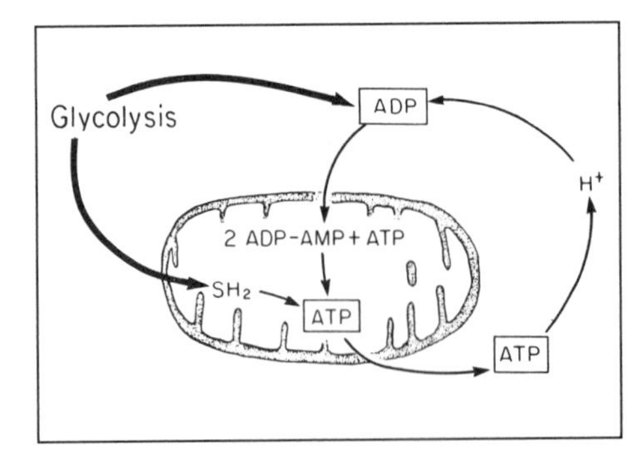

In this slide, another unit of information is added, which increases the overall size of the original drawing. This results in increased reduction of the picture. However, thicker arrows and larger labels on the new information clarify and emphasize the new addition.

SLIDE 3

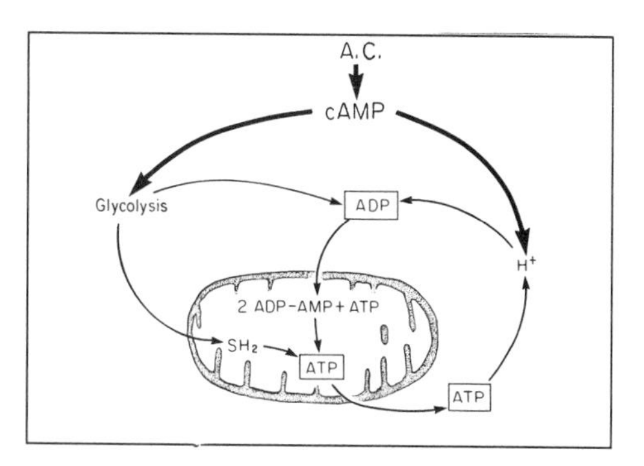

The addition of more information increases complexity and reduction. However, because the viewer now has a background for the newest complicating factor, attention is focused on the new addition with its thicker lines and larger labels.

Another reason for keeping slides as simple as possible is that they must be legible from the back of the auditorium. Although slides projected onto a large screen may seem gigantic from the front row, they appear much smaller in the back row. The previous complicated slide might not be legible except for the enlarged new labels.

SLIDE FORMAT

The outside dimensions of slides that fit into the standard cassette are 2 × 2 inches. Within these 4 square inches is a mat that determines the shape of the slide on the screen. Because slides are usually made with 35 mm film, the format is controlled by this film size.

Slides have a limited number of options for format or shape. The format of the previous figures is the most common. Notice that the shape is more horizontal than it is vertical or square. This is an effective shape for slides and can be easily seen from the back row.

HORIZONTAL FORMAT

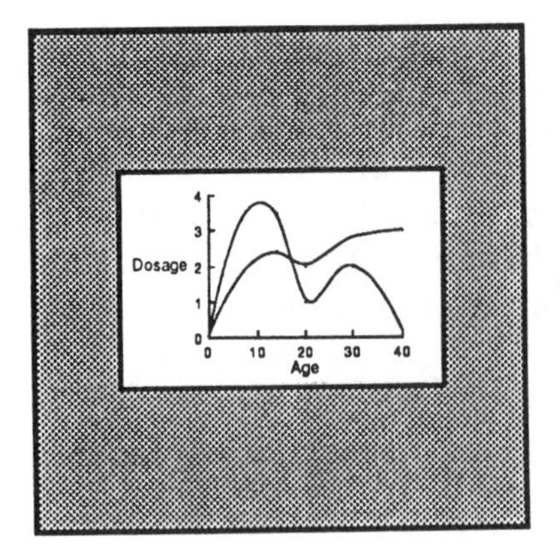

This format and the following format examples are shown actual size. A slide with a horizontal format occupies the middle of the screen. It is at eye level and is easy to read.

There can be problems with the same 2 × 2-inch slide format redrawn vertically.

VERTICAL FORMAT

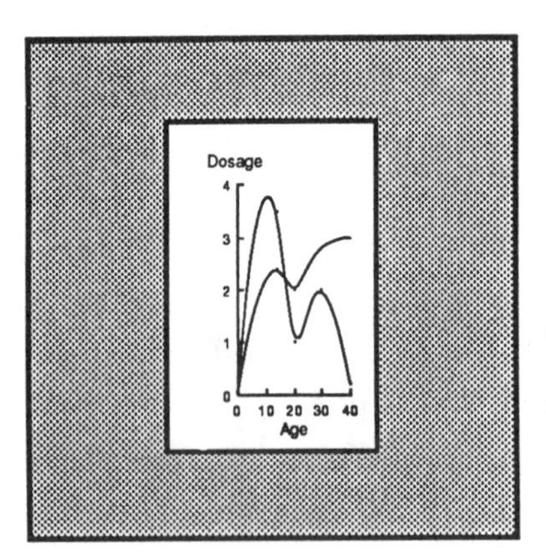

This slide format accommodates a vertical figure nicely. However, from the back row, the heads of people in front may hide the bottom part of this graph. Because a slide screen reaches to the floor does not mean that all that space is usable.

Try to design horizontal figures for slides.

SQUARE FORMAT

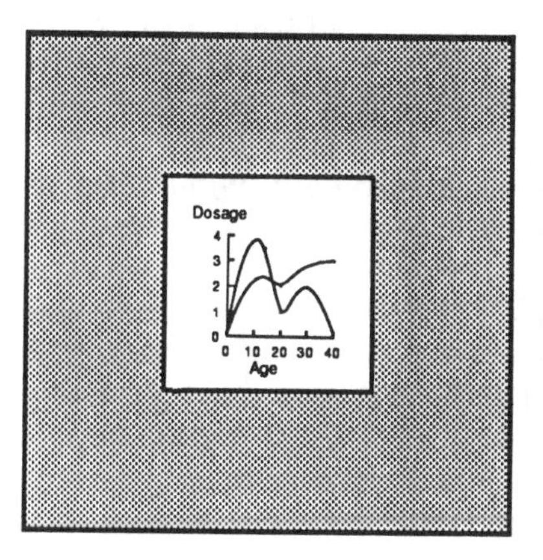

A square format results in a smaller slide, which occupies less of the screen. The surrounding mat takes up more of the space. This square format is advantageous for a square figure in eliminating wasted space on the sides.

To show a square figure to larger advantage, a super slide format may work better.

SUPER SLIDE

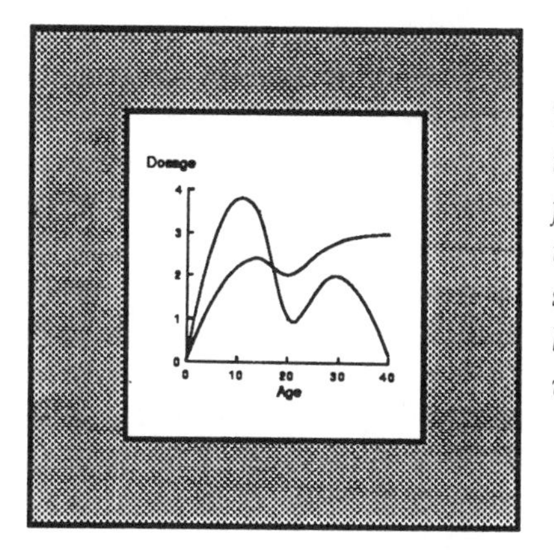

The surrounding mat of a super slide is smaller than the square format, resulting in a slide that fills more of the screen. However, the length of this format is the same as that of the vertical slide, and the bottom part of this format may not be seen easily from the back. Also 35 mm film will not fit this format.

Knowledge of the room or auditorium in which your slides will be shown is helpful in planning for vertical or super slides.

LABELS FOR SLIDES

In general, effective labels for slides are briefer and larger than those for publication. Plain or thin labels are more readable and in better proportion to the rest of the figure than bold thick letters. Bold labeling should be saved for special emphasis. All labels should be horizontal for easy reading.

Plan figures to accommodate horizontal labeling.

There are ways to arrange even long labels horizontally without excessive reduction. Long labels may be divided into several lines and aligned horizontally; or the figure may be designed to leave room for horizontal labels, or labels may be positioned at the top of the *y*-axis.

The width of a figure determines its reduction in horizontally aligned slides.

LONG *Y*-AXIS LABELS

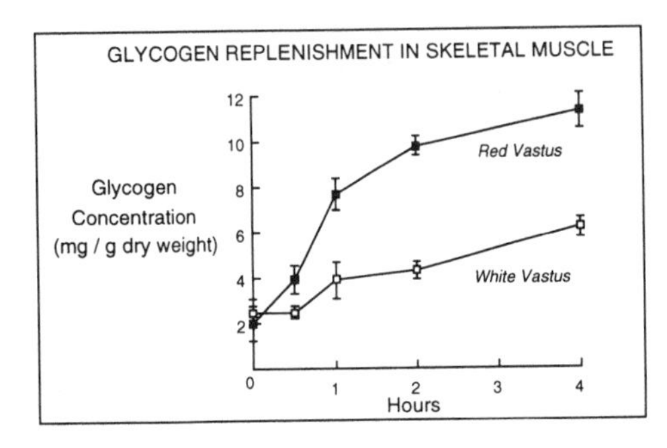

Long y-axis labels in this slide have been aligned horizontally. This is easy to read, but compare the width of this figure to the figure below.

DIVIDED *Y*-AXIS LABELS

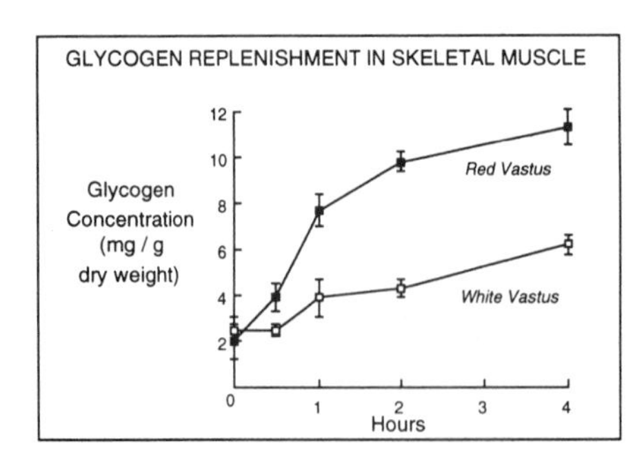

The label describing the units has been divided, so that "Concentration" is now the longest word. Compare the width of this figure to the one above. It is slightly less wide and will make some difference in the figure reduction.

Decreased Graph Width

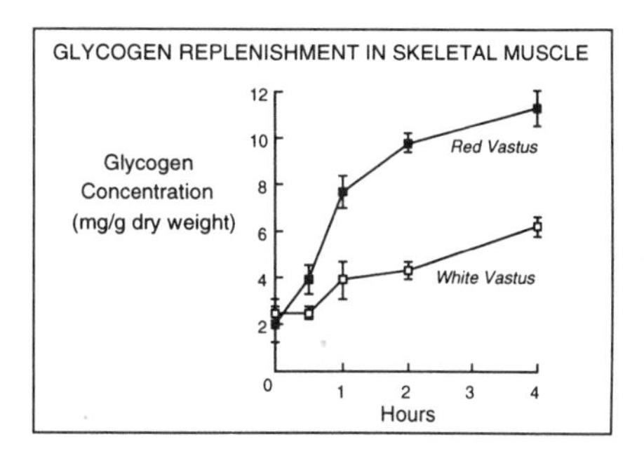

Another solution for long labels is to reduce the width of the graph itself. This leaves room for fairly long horizontal labels. The width is slightly smaller than the previous graph.

Label Above *Y*-Axis

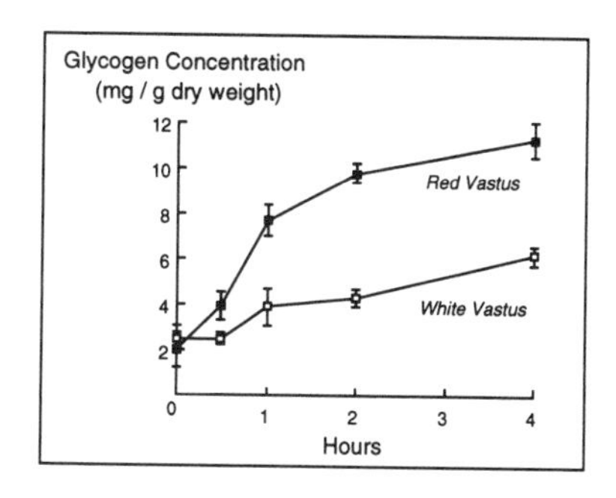

The y-axis label has been positioned above the axis line here. This reduces the width to the extent that the format is almost square. The title has been eliminated too avoid crowed labels at the top of the graph and to avoid an increase in height of the graph.

Although a title may be an asset to a slide, it is not always necessary. The speaker's words may be enough to identify the slide. Also, series of slides that are similar except for one variable may be titled quite briefly with only the name of the variable.

A title for a slide is a brief but informative summary of information. It should be concise and should not repeat what is contained in the axis labels.

LONG TITLE

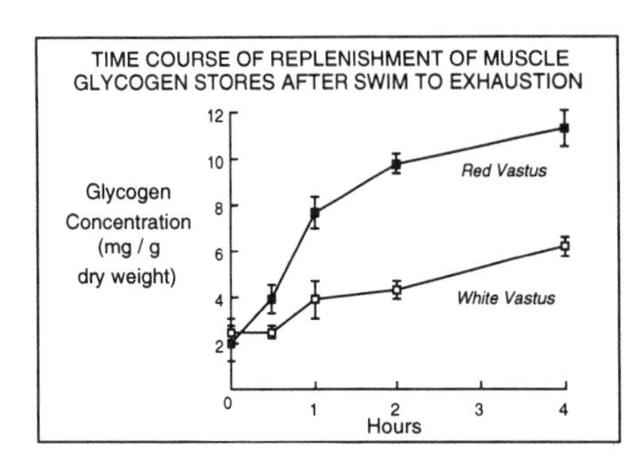

CONCISE TITLE

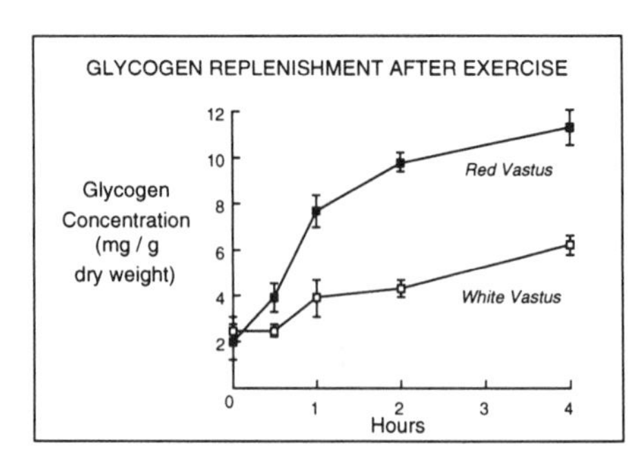

The length of the title on the left is daunting. The "time course" is already stated in hours, "glycogen" is clearly labeled on the y-axis, and "muscle" is labeled red and white on the curves. The only new information in this long title is "exercise." The figure on the right has a shorter and less redundant title, which reduces label clutter and is clearer and easier to read.

For slides, reduce the title information to one line.

COLOR

A good black and white slide is classical and elegant and suffices in almost all situations. However, a color slide can charm an audience and provide variety. More than that, it can be an excellent explanatory tool. *Color keys,* used throughout a talk to denote specific entities, are an easy and pleasurable way to follow the flow of information. They can unify as well as clarify. Color in molecular models is used as a key and is often essential for clarity.

Spot color in otherwise black and white drawings is useful to accentuate or emphasize. In using color as an accent, stick to primary, bright, and clear colors for the best effect. Too many colors may be distracting, and areas of color that are too small may not show up at all.

Full color, as in photomicrographs or full color drawings, may be essential for purposes of identification. It may be a way to capture the audience's attention. The negative of a color photomicrograph may be used as a slide with little loss of brilliance. A positive transparency must be made of a color drawing with an attendant loss of brilliance. Use of strong primary colors on a neutral, rather than white, background is effective for a slide.

Photographing color drawings for use as slides is not easy, as the photographic exposure must be exactly right to avoid an overdark or washed out projection image. A color slide that looks satisfactory on a slide viewer may project poorly. Try out each slide in a suitably sized room.

Color may be added to negative slides with dyes or color tapes. Color background slides may be made from the black and white original. In this kind of slide, the black lines of the original figure appear as white. If these lines are very thin, they are hard to see. Plan for this kind of slide by making lines and labels thicker. Avoid using drawings with complex lines and stipple for this process because their tonality is reversed, causing a distracting negative effect.

Color slides should be used with discretion, not just because they are more costly and time-consuming than black and white slides, but because they may not serve the purpose well. Color can be distracting as well as informative. Have a good reason for using color, not just availability or looks. The purpose of the slide is not to dazzle but to inform.

Word Slides

Many lecturers use word slides as an outline to follow in their talk. This can be helpful to the audience if the terms used are unfamiliar and complicated or if the speaker has a pronounced accent. A word slide also helps to emphasize information such as conclusions.

However, if the speaker merely reads to the audience what is on the word slide, the presentation will be boring for the audience, who can read the words faster than the speaker can say them. There is also danger of losing rapport with the audience when the focus is continually on the screen.

A word slide should contain no more than five short statements. Information should be simplified to the point of being skeletal. It is up to the speaker to fill in the gaps.

PARAGRAPHS

EMBRYOGENESIS CAN OCCUR WITHOUT SURROUNDING MATERNAL TISSUE

Somatic cells from a variety of vegetative and reproductive tissues can undergo embryogenesis in culture, leading to fertile plant production. Also gene expression programs appear to be similar in somatic and zygotic embryos.

Embryo-like structures leading to plantlets can form directly from the attached leaves of some plants.

Zygotes produced by fertilizing egg cells in vitro undergo embryogenesis in culture and give rise to flower-producing plants.

There may be a barrier between the inner ovule cell layer and the embryo sac that prevents the transfer of material directly between compartments.

SIMPLIFIED AND NUMBERED

EGG CELL FORMATION

1. Somatic cells can undergo embryogenesis in culture and produce fertile plants.
2. Embryo-like structures can form directly from the attached leaves of some plants.
3. Zygotes undergo embryogenesis in culture.
4. A barrier may exist between the inner ovule cell layer and the embryo sac.

Paragraphs present dense type with no indications of priorities. The slide on the left is too much and too small for the audience. On the right, information was cut, and statements are numbered and separated. (Information in these paragraphs from Goldberg RB, de Paiva G, Yadegari R. Plant embryogenesis: zygote to seed. Science 266: 605–614, 1994.)

SLIDES FROM BOOKS AND JOURNALS

In the course of being printed, figures in books and journals have undergone several processes of degradation, i.e., photographs made of photographs. Even in the best quality reproduction, there is a loss of sharpness, a possible unevenness of tone, and in the case of continuous-tone figures, a screening into small dots. When projected, flaws become more apparent.

Labeling suitable for a published figure is often too small to be read easily in a slide. The caption that accompanies a figure is most certainly too small to be read and much too long to be included in a slide. Also the shape of the printed figure may not fit the slide format well.

In addition, most often such a figure taken from the printed page does not completely fit the purpose. It contains more information than is needed. How often have you heard a lecturer say, "Don't pay any attention to this part of the slide"? If all the information on a slide is not valuable to the audience, *leave it out.* There are ways to adapt the printed figure to a slide, so do not impose this burden of translation on the audience.

With scissors, tape, white correction paint, photocopying machine, and minimal time and effort, a printed figure may be transformed into an adequate slide.

JOURNAL FIGURE

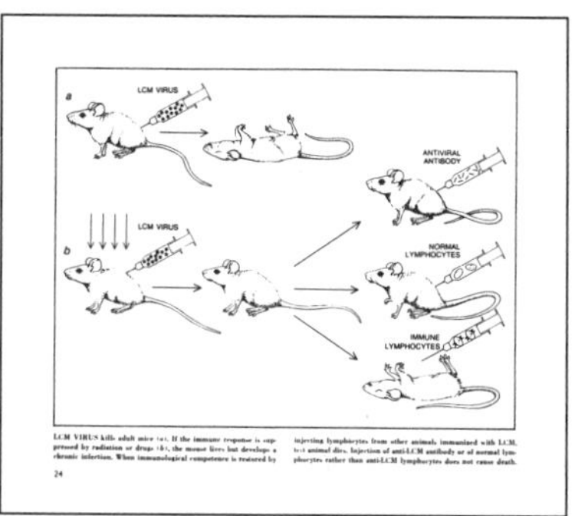

This figure from "How the immune response to a virus can cause disease" by A.L. Hotkins and H. Koprowski (Scientific American, 228(1):22–31, 1973.) is an excellent journal figure, which serves the authors' purpose nicely. However, it contains too much information for one slide, the legend is illegible and unnecessary, and it does not fit the slide format well.

SLIDE FIGURE

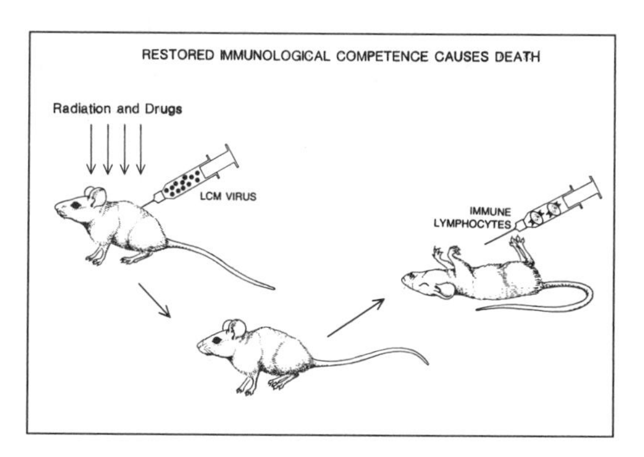

One pathway has been selected for a slide. The selected parts were cut out and glued to a white background. Additional labels, such as the title and explanation for the arrows were added. The layout was designed for minimum reduction in the slide format.

Enlargement may magnify printing flaws. Notice, also, that the photocopy machine thickens lines. This is noticeable in the labeling, where parts of letters fill and blur together. However, this edited version is simpler and more succinct and effective than a straight copy of the original figure.

OVERHEAD TRANSPARENCIES

An alternative to using slides is the overhead transparency. These are figures done on transparent film and placed on an overhead projector, which transfers the figure by means of lens and light to a screen, wall, or blackboard.

Some advantages of overhead transparencies are

- They may be used in a fully lighted room.
- The speaker faces the audience, allowing better audience contact.
- Additions may be made directly on the film by marking pen or by previously prepared overlay sheets.
- They are inexpensive to make.
- They may be made quickly, especially by using a copy machine or computer with compatible transparent plastic sheets.
- Color is an easy option using color markers, color transparent film, and transparent color tapes and press-on letters.

Overhead projectors are usually available in classrooms and lecture halls and can be effective in both small and large settings. It takes little time to learn to set up and operate the projector, so that those who have never used this method of projection could easily experiment with it. Some speakers find they are more comfortable with this direct and simple way of presenting figures.

The best overhead transparencies, like slides, have limited information and large and succinct labels. The buildup of information can be dynamic when the speaker draws directly on the transparency being projected or adds transparent overlays.

MAKING SLIDES AND OVERHEADS

Technology is available that helps to produce professional-looking slides and overheads quickly and easily:

- Scanners with photo design software, such as Adobe Photoshop™, are useful for copying from books, journals or even film to software that allows selection, enlargement, and graphical additions and deletions.
- Presentation software, such as PowerPoint® or Persuasion®, has format options for word slides, equations, and flow and circular charts. Graphs can be made or imported to these programs. Drawings and diagrams may be imported or scanned into these programs.
- Color or black and white are options. Transparencies for overhead projectors can be made on black and white or color printers. Files of figures can be transferred to film recorders. Or disks containing the figures can be taken to print and copy services to be made into slides.

For information on presentation software, call

Photoshop™	Adobe Systems Inc.: (800) 642-3623 or (800) 833-6687
PowerPoint®	Microsoft: (800) 426-9400
Persuasion®	Adobe Systems Inc.: (800) 685-3617

9 POSTERS

A poster at a scientific meeting is an enlarged graphic display containing a title, the authors' names and institution, and text and figures explaining research. This form of presentation was developed as a way to handle the increased size of meetings and the growing number of presenters. Where formerly all information was conveyed in 15 minute talks, now there is a choice of talk or poster or posters only. This method of presentation has advantages and limitations.

Some of the *advantages* of posters are

- They can be studied at leisure or quickly scanned.
- They offer personal contact with interested viewers.
- They can be seen as a whole entity.
- They can be more informative than a talk.
- It is a *visual medium* and excellent for illustrations.

Some of the *limitations* of posters are

- The audience is not captive but must be attracted to the presentation.
- The viewer is not comfortably seated.
- Space is limited, so the poster must be selective.
- Text and figures must be large enough to be seen from a distance of 3 to 4 feet.
- Posters take more time to prepare and cost more money than slides.

PLAN THE POSTER

Planning is the most time-consuming and crucial part of the poster. The *amount of information* to be included is the primary consideration.

Unfortunately, many presenters use the poster as an enlarged journal paper.

THE JOURNAL PAPER PRESENTED AS A POSTER

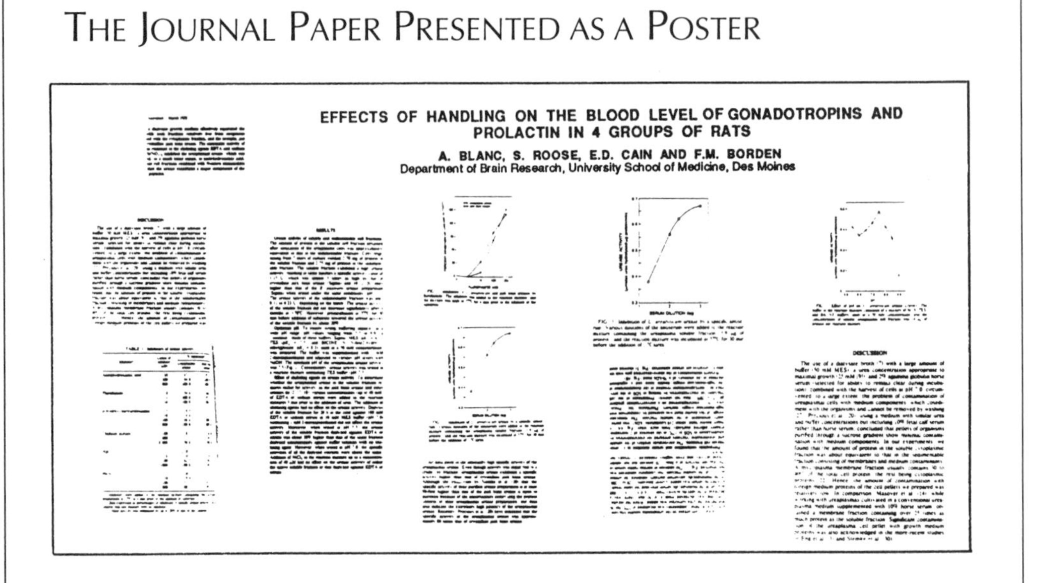

EFFECTS OF HANDLING ON THE BLOOD LEVEL OF GONADOTROPINS AND PROLACTIN IN 4 GROUPS OF RATS

A. BLANC, S. ROOSE, E.D. CAIN AND F.M. BORDEN

Department of Brain Research, University School of Medicine, Des Moines

The text in this poster is written as if it were for publication. Even if it were large enough to be read easily, the sheer amount is overwhelming and discouraging. None but the most determined viewer will stand for the time required to read through all this.

Sadly, too many scientists think an "information-packed" poster conveys the impression of highly productive research. It does not. The resulting crowded and confusing presentation obscures any central ideas and conveys the message that the research may be equally unclear.

Limit your information. Get to the heart of the matter and leave out the rest. In planning what you want to say, pick out no more than three points that you think are most important and focus on them. If you can get across even one point clearly and quickly to your viewer, your poster is successful. Remember that you are there to answer questions and fill in details. If necessary, make use of printed summaries for viewers who are interested in more detailed information.

Poster Instructions

Poster instructions (provided by meeting organizers) have information that is vital to planning, such as size, location, length of time for viewing, information to be included, and often layout suggestions.

For planning, it is essential to know the poster size. Do not take it for granted that all posters are 6 feet wide by 4 feet tall (2 × 1.33 meters). Many are 8 feet wide by 4 feet tall; some are 4 × 4; others may be taller than they are wide. The size of the poster is the first indication of how much to limit the information. The size of the poster also determines the layout.

POSTER LOCATION

Poster sessions are located in hotel halls, in hotel rooms, or in cavernous convention halls, as in the figure above. The noise level may be high and the lighting level low, neither situation being conducive to thoughtful leisurely perusal of a poster. This is another reason to limit the information.

The length of time for viewing varies. An hour is the minimum, 23 hours the maximum. If the poster will be up for more than 1 hour, the presenter is not usually expected to be there the whole time. This means that the essential points should be clear and easily understood without personal explanations.

Information in the poster includes a banner with the title, the authors' names, and the institution. Sometimes the poster number and the abstract must be added to the banner. The banner is always positioned at the top of the poster, and instructions usually specify that letter height be 1 inch (2.5 cm) or larger.

Poster Title

Before sending in the abstract, it is worthwhile to consider how the title will appear on the poster. Lengthy poster titles will discourage the viewer. The first thing that a viewer will see is the title, so it should be brief, informative, and interesting. If a title is enticing or provocative as well, it will attract the viewer's interest.

The following title is lengthy but reasonably informative:

MECHANISM OF AIRWAY CONSTRICTION AND SECRETION
EVOKED BY LARYNGEAL ADMINISTRATION OF SO_2 IN DOGS

However, the following version states the conclusion and is shorter and more informative:

EVIDENCE THAT REFLEX EFFECTS OF SO_2 ARE MEDIATED BY
AFFERENT ENDINGS IN THE UPPER AIRWAY

But the next version is even shorter and, because it is a question, attracts the viewer's attention:

ARE REFLEX EFFECTS OF SO_2 MEDIATED
BY AFFERENT ENDINGS IN THE UPPER AIRWAY?

Even shorter, although perhaps too terse, is the following:

HOW DOES SO_2 AFFECT THE UPPER AIRWAY?

The first title, when enlarged to 1-inch-high letters, will stretch across 5 to 6 feet of width (2 meters or less). With the addition of names of authors and institution, the height of the title will be at least 6 inches (6.66 centimeters). A long title, large enough to be read from 20 feet (6.66 meters), will overwhelm the poster.

Abstract

Many meetings specify that abstracts be posted. However, all the information in the abstract should already be on the poster in a large, easy to read, organized form. An enlarged abstract will add nothing to the poster.

POSTER WITH ABSTRACT

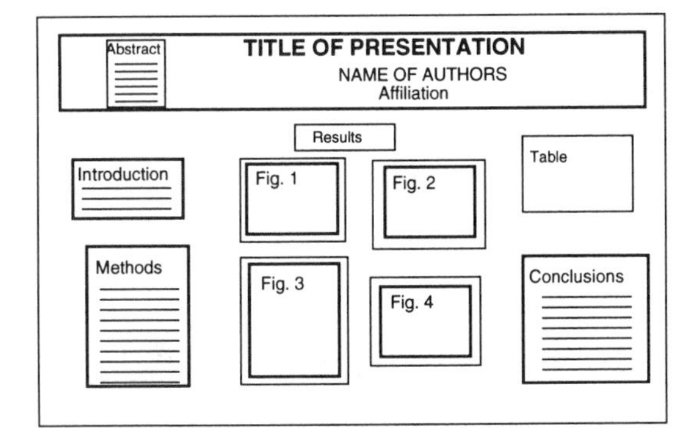

This is a suggested layout that might be included with the poster instructions. The space allotted to the abstract is in the range of 9 × 4 inches. Because it is posted at the top of the poster, near the title, and cannot be adequately enlarged, it will not be readable and would be better left out.

The meeting program will usually contain a copy of the abstract, which can be read at leisure. It is a waste of time and money to type and enlarge the abstract. Either leave it out or use a photocopy of it.

Rough Layout

To visualize size and position when planning the poster, sketch a rough plan. This will give a general idea of how much text and how many figures should be included in the poster. This sketch should show approximate positioning and size of figures and text. The square, vertical, or horizontal shapes of your figures and text begin to evolve in a rough layout. Experiment with a number of rough layouts. Try different sizes and positions before deciding which layout suits the information better.

ROUGH SKETCH

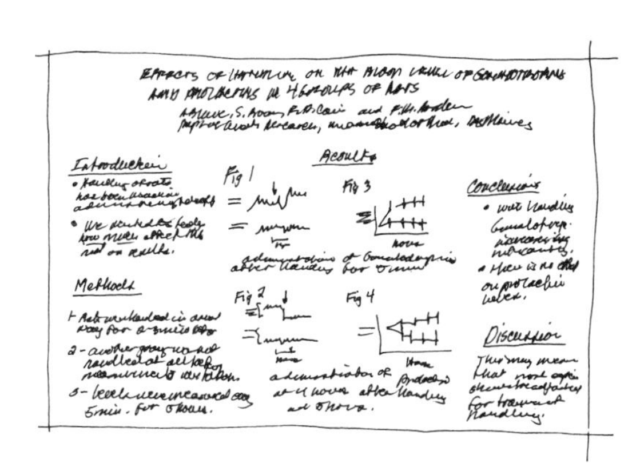

The poster dimensions are approximately 4 × 6 feet (2 × 1.33 meters), and information is organized into broad topics. The shapes of pre-existing figures are roughly drawn.

In the rough sketch, start to distill your ideas into only the most important points. Decide which one or two points you want the viewer to understand.

Do not try to increase the number of figures or amount of text by making text and figures smaller or by squeezing them together. If the poster is hard to read or cluttered, it will be ineffective. Look again at the information to see what further cuts may be made. Be ruthless and single-minded. *Never lose sight of the few points you want to emphasize.*

> It takes intelligence, even brilliance, to condense and focus information into a clear, simple presentation that will be read and remembered. Ignorance and arrogance are shown in a crowded, complicated, hard-to-read poster.

POSTER TEXT

In writing papers, thoughts are expressed in paragraphs full of sentences that are often long and complex. This verbose kind of writing for a poster causes the viewer to struggle to find the core idea. Plan the poster text in short, simple, separate statements. This allows the viewer to scan the text quickly and easily for the important points.

PARAGRAPH

Low concentrations of SO_2 cause bronchoconstriction in asthmatic patients. Since low concentrations of SO_2 may be totally absorbed in the upper airways and since the upper airways appear to be very sensitive to SO_2, we have explored the possibility that SO_2 evokes reflex effects by engaging afferent nerves in the upper airways.

SEPARATE STATEMENTS

Bronchoconstriction in asthmatic patients is caused by SO_2 in low concentrations.

Upper airways are sensitive to and totally absorb low concentrations of SO_2.

We explored the possibility that SO_2 engages afferent nerves in the upper airways.

The text on the right can be understood much more quickly than the paragraph on the left because the statements are separated and simplified. Key words such as "bronchoconstriction" should be put first. Not only does this make it easier to scan text quickly, but it emphasizes important information.

Text for a poster usually includes an introduction or background, methods, results, and conclusions. Because the type must be large, you will have to cut down the amount of information. Text should also be titled to make the different sections under discussion immediately clear.

TEXT TITLE

INTRODUCTION

SO_2 in low concentrations causes bronchoconstriction in asthmatic patients.

Upper airways are sensitive to and totally absorb low concentrations of SO_2.

We explored the possibility that SO_2 engages afferent nerves in the upper airways.

For quick identification, text should have a title. If each section of text is titled, the flow of information on the poster becomes more apparent. Here the first statement emphasizes low concentrations of SO_2.

Text should be about one-quarter inch high (24 point or larger). On the following page is an example of text size that will give you an idea of how much information may be included in a given space. The title is 36 point bold; the text is 30 point plain.

Try viewing both of these pages from 3 feet. Even those with perfect eyesight will have to admit that the large lettering on the following page can be read much more quickly and easily than the small print.

Remember that many in your audience have reached the “bifocal age” and have difficulties focusing on small type.

This text can be read without strain from 3 to 5 feet. Viewers will be drawn by its simplicity and size to scan the text quickly for core ideas.

The *font* in the sample text is Helvetica: a simple, clean, and legible type face. It is a sans serif font, which means that it lacks the fine lines attached to the extremities of the letters. It is also proportionately spaced (some letters are closer together than others), which is easier on the eyes. The title is bold (thick),

TEXT SIZE

INTRODUCTION

Bronchoconstriction in asthmatic patients is caused by SO_2 in low concentrations.

Upper airways are sensitive to and totally absorb low concentrations of SO_2.

We explored the possibility that SO_2 engages afferent nerves in the upper airways.

Text of this size may come as a shock to those who persist in thinking of posters as enlarged papers. Most posters that you will see in meetings have text that is too small and too complex.

and the body of the text is plain (thin). This is an excellent font to use for posters because it is easy to read, does not distract from the information, and yet is attractive.

Other sans serif fonts are Univers and Avant Garde. They are shown below in 30-point size to compare with the text on the previous page.

FONT VARIATION

UNIVERS

We explored the possibility that SO_2 engages afferent nerves in the upper airways.

AVANT GARDE

We explored the possibility that SO_2 engages afferent nerves in the upper airways.

Univers and Avant Garde are proportionately spaced type. Avant Garde is thinner lined than Univers or Helvetica. Notice the differences in letter shapes and overall looks. All are readable; all are conventionally shaped and will not distract from the information they describe.

If a serif type such as Times is used, it should be used consistently throughout the poster. The text of this book is printed in Times, a serif type with extra lines on the letters. A serif type is said to be easier to read in extensive, small type, but in the large, economical text of a good poster, there is no difference in ease of reading. A serif type can be distracting, especially in a large title.

Save the use of bold labels for special emphasis, such as the title. Text that is all bold becomes strident and tiring to read. It can also overshadow the figures.

In addition to cutting information and making figures and text large, use simple clear language. Avoid abbreviations and jargon. Avoid tables if possible. If absolutely necessary, make them brief or put them into a handout.

Entice the viewer with clarity, simplicity, and pictures.

DOES THE POSTER DISCOURAGE THE VIEWER?

FIGURES

Figures are more impressive than text on a poster. The poster medium is made for pictures. Because we learn much more quickly and easily from pictures than from words, use figures to tell the story and plan the poster around the figures.

Drawings will attract the viewer. They hold the viewer's attention and communicate vividly and memorably. They are especially effective in describing methods.

METHODS DRAWING

FIGURE 1

EXPERIMENTAL PREPARATION: VENTRAL VIEW

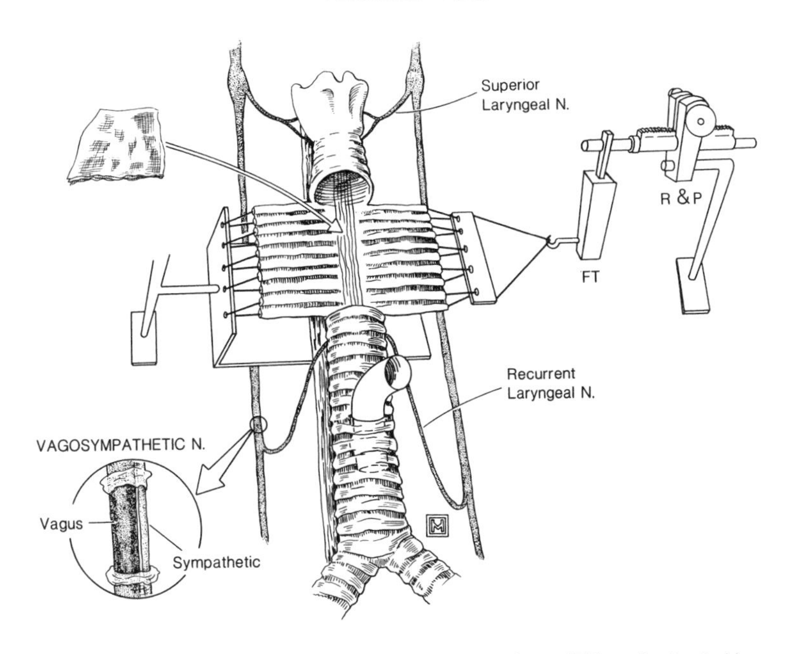

The tracheal segment is connected to a force transducer (FT) and adjusted by rack and pinion (R & P). The metal plate under the cut trachea prevents disturbance of the laryngeal nerve branches. The sympathetic nerve is isolated and surrounded by stimulating electrodes, and contraction is induced by acetycholine applied to gauze.

This drawing shows an experimental preparation of the trachea. Although it was designed for a paper, it could be enlarged to make an effective poster figure that could take the place of a long description of methods. A brief explanation of abbreviations and procedure is effectively shown directly under the drawing.

Graphs must be large. A standard 8.5 × 11 inches is a minimum size. Axis labels should not be less than 24 point. Below is a graph enlarged to a size that is easy to see from 3 to 4 feet. Because of the necessarily large size, information on graphs should be limited and labels should be short.

MINIMUM FIGURE ENLARGEMENT

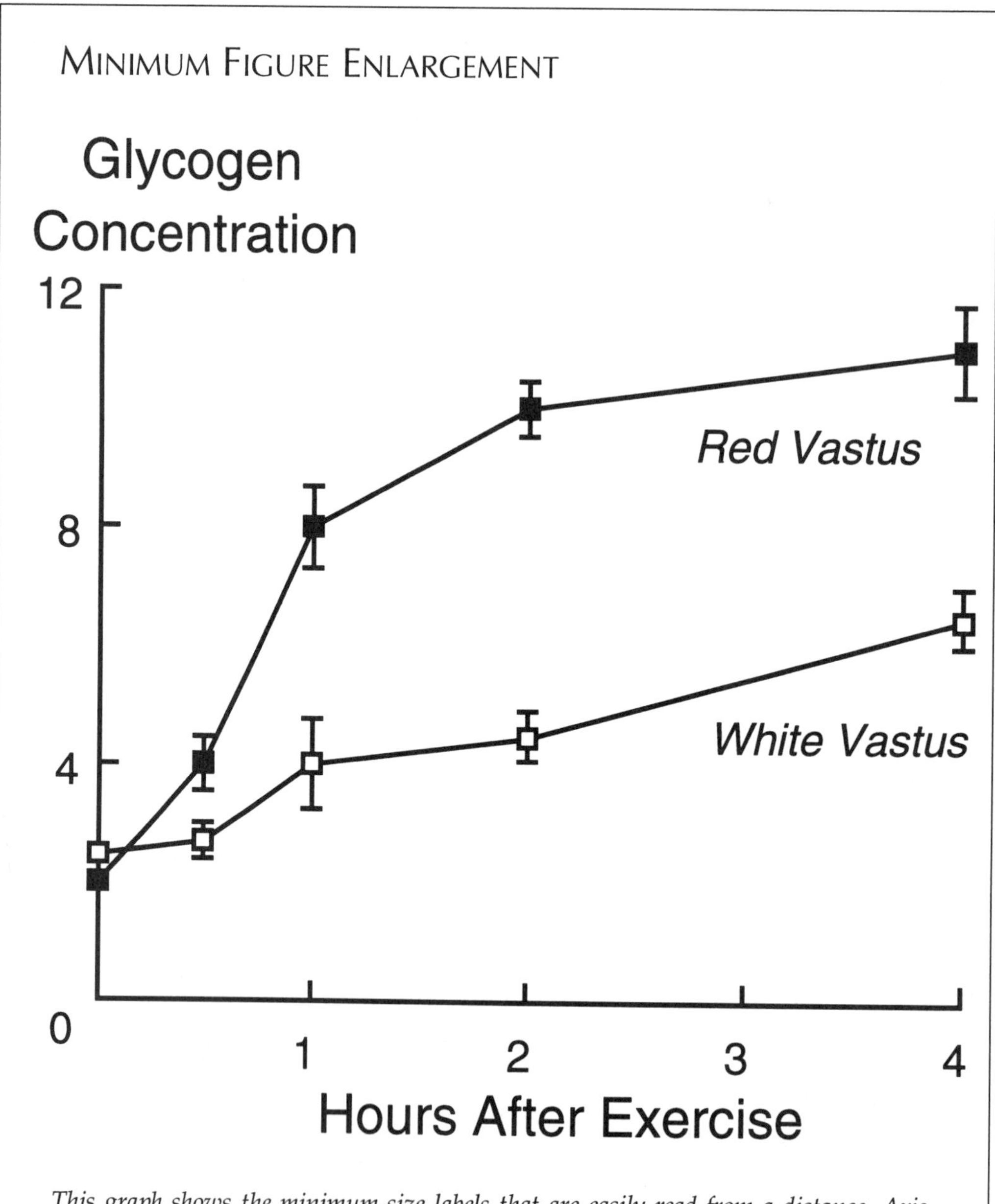

This graph shows the minimum-size labels that are easily read from a distance. Axis labels are 24 point; all other labels are 18 point. This figure would be even better if it were turned sideways on the 8.5 × 11 inch page and enlarged still more.

Figures should be numbered prominently for ease of identification and flow of information throughout the poster. If the figure is referred to in the text, it may then be referred to by number. If the figure number reference in the text is made prominent by being bold, all upper case, or larger, it facilitates cross-reference of text to figure, figure to text.

A concise explanation under the figure is helpful for clarification. In a short statement, details that were left out of the text may be included.

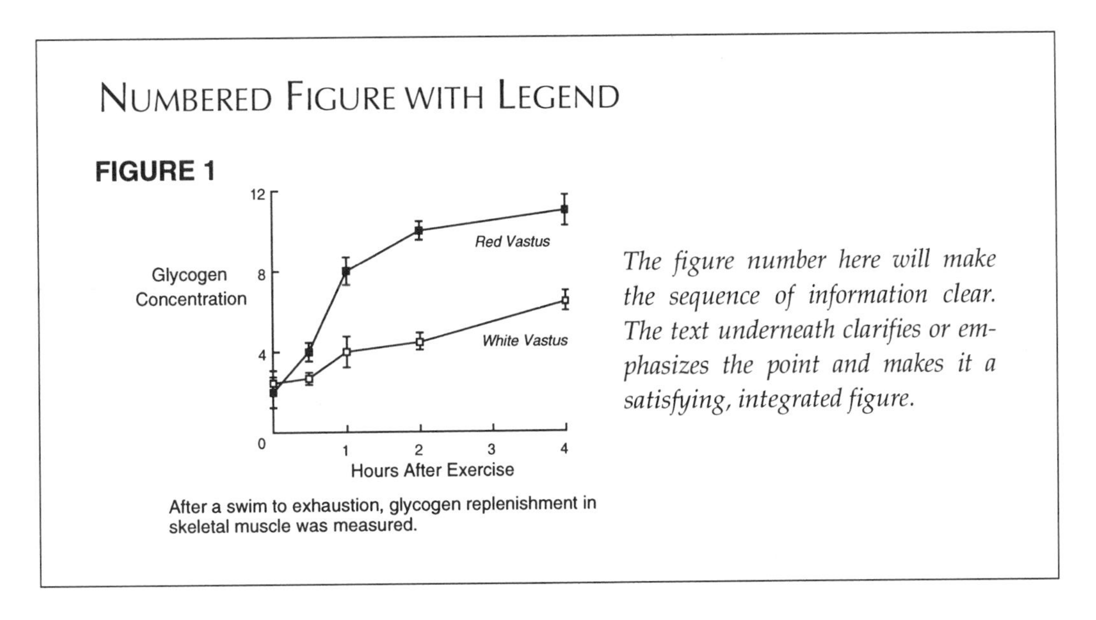

After a swim to exhaustion, glycogen replenishment in skeletal muscle was measured.

POSTER LAYOUT

A poster should be organized so that the progression of information is clear, keeping in mind that we read from top to bottom and left to right. Important information should be at eye level. The top of the poster will contain the title, which is usually read on the approach to the poster. The 2 feet of space below the title is at eye level for most and is the area where information is most easily read.

An accurately scaled, penciled layout will help in arranging the information. Make a grid in inches instead of feet or use graph paper. This serves as a blueprint for relative sizes, juxtapositions, and flow of information.

Plan the poster to be read in sections from left to right and top to bottom. For a 6-foot-wide poster, it is best to divide the space into three or four sections. By doing this, each section can be read while standing in one place. The viewer need only move to the right to read the next section. This is especially practical when

READABLE SPACE

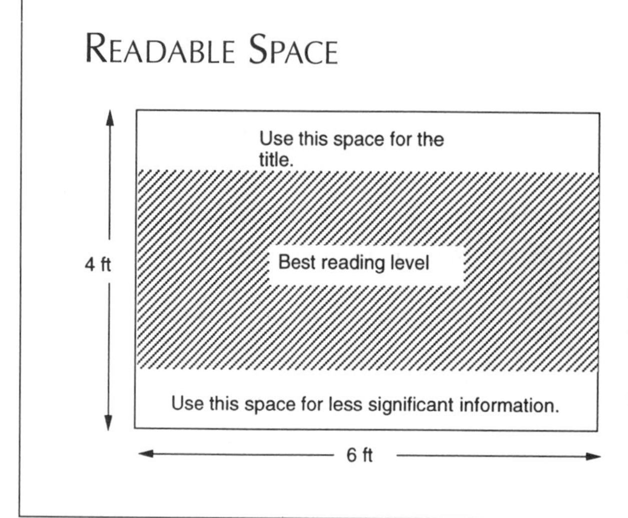

Notice that some of the available poster space falls outside the best reading level. This is another reason for strictly limiting the amount of information. Plan for space around your text and figures. Do not try to cram the whole space with information.

there are many people milling around the poster, it makes it possible for several people to read the poster at the same time.

Before the meeting, arrange the poster as it will be mounted to get some idea of how it will look. Check for consistency of style, terminology, and symbolism. Show it to your colleagues and to scientists outside of your expertise. Ask them and yourself if the poster presents a clear, simple, cohesive message.

Many laboratories and groups have poster practice sessions to which colleagues are invited for a discussion and critique. Time should be allowed to make changes before the meeting.

POSTER DIVISIONS

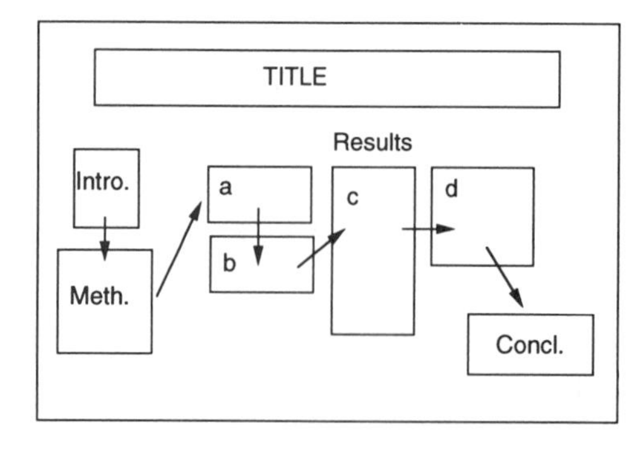

Sections are delineated by grouping of figures and texts and by leaving space between the groups. The arrows indicate one way the eye naturally moves through the poster.

POSTER PRODUCTION

There are many ways to prepare and assemble the poster materials, from time-saving, simple, and inexpensive to time-consuming, sophisticated, and expensive. Although your own resources may be a limiting factor, the one thing you should not stint on is your time. Even if the poster is produced by a commercial design studio, the time you spend carefully planning and organizing the information will determine its success. The time you spend consulting with the artist and photographer will make a difference. If you produce your own poster, plan to spend several days on it.

Black and White Enlargement

The size of *photographic* enlargements is limited only by available photographic paper size. Photographic enlargement is the most faithful to the original. For posters, such enlargements should be printed on a matte surface paper to avoid glare and reflection.

A *PMT* is a photomechanical transfer in black and white. It is printed on a matte finish paper and is useful for enlargements up to 18 × 24 inches. Because it requires no negative, it is less expensive than a photographic enlargement. An 18 × 24 inch-print costs about $20.

Enlargements by *copying machines* are generally limited to 11 × 17 inches. One-ply bond paper is usually used, although other finer one-ply papers, such as plate finish bristol, may sometimes be substituted. At 10¢ per copy, this is the least expensive kind of enlargement.

Enlargements may be made by *computer* and printed on a laser printer. Although maximum paper size is 8.5 × 14 inches, paper may be spliced together for any size. Print quality is good, and for anyone with access to a computer/laser printer, this is a quick, flexible, and inexpensive way to produce title, text, and figures.

Photo murals are enlargements of the whole poster. Individual parts of the poster are mounted together in half or third size and then enlarged to the final size on one large sheet of paper. This is rolled up for transport to the meeting, then unrolled and tacked to the poster surface.

Paper for this process comes in rolls. Two processes are available: the stat/negative process and the photocopy process. The maximum width of paper for the stat/negative is 42 inches, slightly less than a 4-foot poster height. The negative costs about $10, and the print itself costs about $5 a square foot, so that a 3.5 × 6-foot blow-up will cost about $115. It is printed on photographic paper.

A photomural done by the photocopy process is printed on light bond paper and undergoes enlargements or enlargements, so the final result may be poor quality. The maximum width of the paper is 36 inches, which is a foot less than the 4-foot poster board. The cost of a 3 × 6-foot reproduction ranges from $18 to $36, depending on the copy service.

Text Production

Typewritten text is not a viable alternative for posters unless it is enlarged. However, to go from typewriter-size text to a reasonable poster size means an enlargement of 400–500%, often with extreme loss of quality.

Text that is *typeset* results in enlargements of excellent quality but entails not only the typesetter's charge but an additional cost of enlargement.

Computer-typed text, printed and enlarged by the laser printer, is the best alternative. The print quality is good, and changes in text and enlargement may be easily made. A word processing program is most flexible for this purpose, although some drawing programs provide good text options.

Trimming and Mounting

Trimming of figures, text and title may be necessary to save space or to splice pages together. This can be done with an accurate paper cutter or sharp knife against a metal ruler.

For the most basic production, the enlarged, separate text figures and title can be taken to the meeting and tacked onto the assigned board according to plan. However, mounting the separate parts on light board adds body, durability, and color.

Material for mounting may be one of the following:

1. Poster board and railroad board are lightweight boards that come in a variety of colors.
2. Mat board is heavier and also available in many colors. It is harder to cut than the lighter boards.
3. Cover stock and construction paper are heavy papers, easy to cut, and available in an array of colors. Although these papers are not as durable as heavier boards, they are easier to transport because they are thinner and weigh less.

To mount the poster to the backing board, use a *spray adhesive* such as 3M Photo Mount™. This coats the paper evenly and is quick and easy to use. The item to be glued is placed face down on newspaper or other disposable material. This prevents excess spray glue from covering the floor or table surface. The back of the item is sprayed, then pressed into position on the backing material.

Rubber cement takes more time to apply and is not as smooth as spray adhesive. Glue Stic™ is handy for small items but also takes time to apply evenly.

Poster Backgrounds

The poster surface provided at most meetings is tan cork board or white composition board. They are neutral to ugly looking. The white background swallows and absorbs the white paper on which your poster parts are printed. In this case, a color background is essential for contrast, emphasis, and clarity.

Often a professionally done poster uses a *modular mat board background* that covers the entire poster space. Figures are mounted on modules of backing board so that when they are tacked to the meeting board, the pieces form a solid colored background. This is time-consuming and requires expertise in accurate positioning of figures and cutting of boards. The result, however, is a consistent background color that frames the poster text and figures and unites them into a cohesive whole. Different color can be used for specific sections of the poster. This is a way to divide or emphasize poster parts.

Another possibility for background color is *Velcro™ material* cut to the poster size and tacked to the meeting board along the edges. When pieces of adhesive loop tape are applied to the backs of the poster parts, they can be quickly mounted on the Velcro-covered board and just as quickly changed around and taken down.

There are also heavy *seamless papers* in 53- and 107-inch widths in many colors. These papers may be cut to poster board size and tacked to the board. This provides a smooth color background for tacking poster items.

Color

Color is a great asset in posters. A color background frames and unifies the poster parts. Color keys used consistently throughout the poster make information easier to follow. In addition, color is pleasing and attractive to the viewer.

If possible, use color backing. Apply color to columns in bar graphs with color markers or color film. Use color paper for arrows; use color transfer letters for short labels; use color photographs and drawings. Just make sure they are large.

Although large photographic color prints are expensive, less expensive color prints can be made using the color photostat process or the color copier:

- The maximum size for a *color photostat* is 11 × 16.5 inches and maximum enlargement is 200%. A color slide may be enlarged to this maximum size. In this process, the range of colors is fairly subtle, and the background will be a neutral color similar to that of a photographic print. Color contrast may be dampened in this process by the light gray background. A maximum-sized stat costs about $25.
- The *color copier* uses a laser printing process, and maximum paper size for printing is 11 × 17 inches. Each copy costs about $4. Color quality in these prints is brighter and more artificial, tending to exaggerate yellow, but the background is the pure white of the bond paper used in this process.

Color printers for *printing computer-generated files* produce brilliant colors. Files must usually be saved in encapsulated PostScript language (EPS), which is built in to most drawing programs, such as MacDraw®, Adobe Illustrator™, and Freehand®. It is possible to transfer non-EPS files as from graphing programs to a draw program. The cost of an 11 × 17-inch print is about $18.

Color photo murals can be printed from EPS files onto computer paper and cost about $175. Because these files are generally large, it is preferable to save them onto a mass storage disk for printing.

With the multimedia software available now, it may not be long before a poster will consist of a laptop computer projecting onto the poster board, continually replaying the text and figures, with sound and even animation built in!

Transport

Most meetings are held away from home and entail plane travel with its attendant baggage and space complications. Consider this when deciding how to produce the poster. For easy transport, the poster should be lightweight and compact enough to take on the plane. Checking the poster with baggage risks losing or damaging it.

A photo mural can be carried rolled into a 4 foot mailing tube.

Posters mounted on large sections of mat board are awkward to carry, but sections may be cut and taped so that they can be folded to a more compact size. If they will not fit into a briefcase or large envelope, at least cover them with heavy wrapping paper to avoid dirt and damage.

Better yet, use a heavy, corrugated mailing box. These come in a 23 by 13 × 2.5-inch size and can be located under "Boxes—Corrugated & Fiber" in the Yellow Pages.

Separate poster items should be planned to fit into a briefcase or heavy box. Velcro™ material can be folded and put into a suitcase or box, and seamless paper backing can be rolled into a mailing tube.

POSTER PURPOSE

Communication of one or two ideas should be the purpose of the poster. The purpose of a poster is *not* vanity or ambition. Unfortunately, in the crowded, noisy, poorly-lit halls for poster sessions, there is often very little communication at all.

Scientists could learn from advertisers. Compare the average scientific poster with the sales exhibits at a scientific meeting. Whatever the agenda of advertisers, they communicate effectively. Advertisers spend large sums of money, thought, and time to present limited information concisely, clearly, and attractively.

Observe and think about what communicates to you, what communicates to your colleagues. If you genuinely desire to teach others through your poster, you will not be afraid to be simple, clear, and creative.

10 Using an Illustrator

Art and science, far from being exclusive, have a long history of symbiotic, fluid, and rich relationships. Vesalius, daVinci, Audubon, and Beatrix Potter are famous personifications of this marriage of science and art.

A medical or scientific illustrator is an expert at making drawings, diagrams, tables, and graphs. Ideally, an illustrator's expertise should be available to all scientists. An artist's way of seeing scientific information and rendering that information into effective visual presentation can be of great benefit to scientists.

Having done the research, the scientist expects details, accuracy, and flexibility of data presentation. The artist, concerned with communicating understanding of a clear message, looks for visual accuracy in communicating the large story the data tell and the visual details that will display that message best. Collaboration between artist and scientist, if only by consultation, is often the preferred way to do figures.

A trained scientific illustrator is more proficient than a scientist in some areas of illustration, and some figures are more efficiently done by an illustrator.

For complex drawings or diagrams, an illustrator is a necessity. For tables, charts, and graphs, an illustrator can save time. An illustrator can help plan and can evaluate and make suggestions for figures done by hand or by computer.

An illustrator's skill in drawing and design can turn an adequate figure into a clearer and more interesting figure.

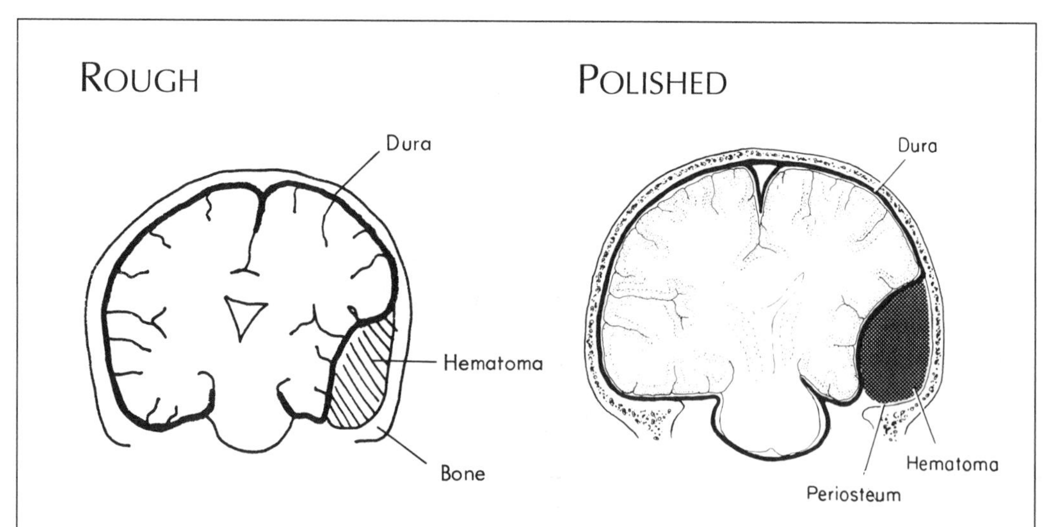

Both of these drawings show blood collected between the dura and periosteum, forming a hematoma. On the left, the general area of the hematoma and rough configuration of the dura are apparent. But the artist's rendering on the right is more exact and contains more detail. It is also visually more interesting.

An illustrator can help communicate by visually synthesizing, clarifying, and simplifying your ideas. An illustrator can help plan the figures by presenting choices and suggesting alternatives. An illustrator can serve as a consultant for figures made on the computer.

COMMUNICATE WITH THE ILLUSTRATOR

Before seeing an illustrator, take time to plan and organize your ideas. Start planning well before the deadline.

- Know what you want to show in the drawing. Decide on priorities.
- Check journal, poster, or meeting instructions for size and final form.
- Check for accuracy of ideas or data. Discuss rough ideas with colleagues or coauthors.

Provide the illustrator with the following information:

- The meaning and significance of the figure.
  The illustrator must understand at least the general nature of the information to suggest ways of presentation.
  Take advantage of an illustrator's expertise by asking for such suggestions.

- The purpose of the figure: paper, slide, or poster.
- The deadline.
- Raw data or rough drafts of information or ideas.
- Any reference material that will be needed.

Be aware of the steps involved in drawing:

- Research and/or on-site sketching.
- Decision about the *technique* to be used. For example:

  Pen and ink reproduces well in journals and is economical to print.

  Carbon dust, pencil, watercolor, or airbrush tone rendering is smooth and lifelike. These continuous-tone techniques usually take more time than pen and ink, and in reproduction, the background is gray. For slides, continuous tone is effective. For journal reproduction, it is more expensive.

  Color paper, color film, color pencil, pastel, watercolor, or airbrush rendering are effective for slides and posters but are expensive for publication.
- Several sketches before the final sketch.
- The final drawing.
- Possible changes anytime in the process.

Budget more time and money for drawings and color work than for black and white graphs. Check all work carefully and allow time for changes, additions, and corrections. Allow 5 working days for photographing of the final artwork.

THE ILLUSTRATOR AND THE COMPUTER

The widespread use of the computer for charts and graphs has been a boon for the illustrator as well as the researcher. The time and tedium involved in hand-done repetitious graphs have been eliminated. Most illustrators can enter the data and plot and print the graph effectively and quickly by computer. An illustrator also brings expertise in design to this exercise, which may make the difference between a good or bad graph.

Almost all computer-drawn graphs require visual editing. Visual editing involves observation of what works visually to communicate information. It involves making the educated choices described in Chapter 6, "Graphs and Software." An illustrator can look at the graph below and suggest simple ways to improve it.

ROUGH COMPUTER GRAPH

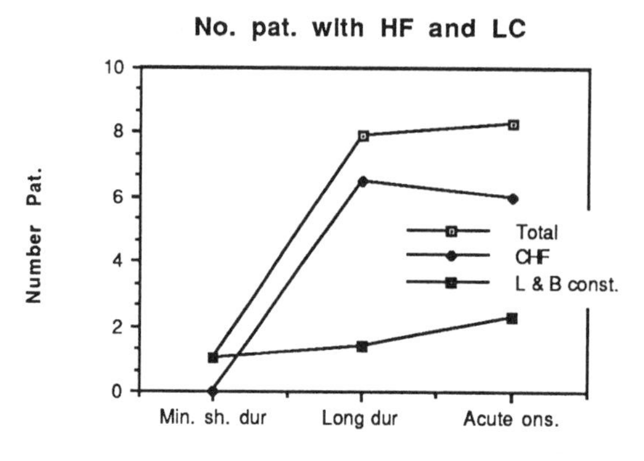

The graph on the left shows the total number of patients who died, broken down into symptoms and kinds of illness. This is an example of information that looks more complicated than it is. Abbreviated labels are unclear; the y-axis is cluttered with numbers and ticks; the frame around the graph serves no purpose. Above all, a line graph is not the best way to show totals with divisions.

POLISHED COMPUTER GRAPH

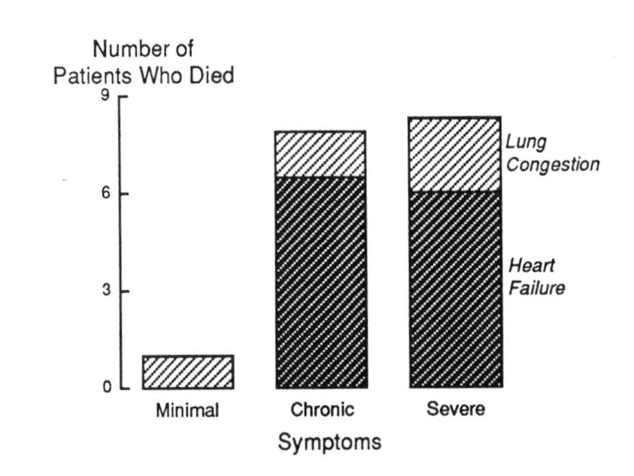

An illustrator could make the changes shown in this graph or could suggest that you make such changes. Here information is shown in a stacked bar graph that shows totals with divisions into the kinds of disease. Labels are written out in simple terms; the y-axis is simplified and the plot frame removed.

THE ILLUSTRATOR AS A RESOURCE

An illustrator can describe and demonstrate materials and their uses and can recommend sources for materials and services. Long experience with the assets and limitations of tools and different media make the illustrator a valuable and practical resource.

An artist can be an antidote to the scientist who is so immersed in the work that what seems simple to him after years of involvement "must be clear to others." An artist often is more useful than a colleague in looking at visual material objectively and in assessing whether this material will be understood outside of the scientist's specialized field.

Above all, an artist has an ability to simplify and get to the point visually. This trait is essential for communicating visual material. It is a trait that is often lacking in scientists who are trained in verbal but not visual skills. The artist is trained to observe details and relationships in the context of a cohesive whole picture, whereas the scientist often gets bogged down by details.

11 Using a Computer

Scientists are finding that, besides being a useful tool for research, computers enable them to do graphs, tables, word slides, and simple diagrams quickly and easily. Personal computers are increasingly affordable and are easy to learn to use. There is excellent software for making illustrations on the computer. The computer-generated examples in this chapter were all done on the Macintosh computer and printed out on a laser printer, but they could also have been produced on an MS-DOS® machine.

Computer-generated diagrams and graphs are certainly within the capabilities of any scientist, and the use of computers by scientists often helps to clarify ideas and to increase visual perception. The use of computers by scientists for figures can be a rewarding and enjoyable experience.

Some assets of computer use for preparing illustrations:

- Speed.
- Ease of use.
- Ease of change.
- Possibilities for experimentation.
- Professional-looking results.

As with any tool, the end product depends on the skill of the user. The multiplicity of choices offered in most programs requires that the user have clear goals and knows what works to communicate well visually.

Some difficulties of computer use for doing illustrations are

- Program limitations.
  A figure may not fit the program parameters and may have to be adapted to what the program can provide.
  Nonstandard figures that do not fit into one program may require transfer to another program.
- Computer speed and ease abet a tendency for rush jobs, often resulting in carelessly planned and executed figures.
- Complicated figures and drawings can take a great deal of time.
- A preoccupation with the medium itself may distract from the main goal: to communicate clearly and simply.

A computer is a tool that takes time, practice, and experimentation to get good results. Excellent scientific figures require awareness of the requisites of good illustrations discussed throughout this book. Do not be dazzled by its speed and versatility. Do not let the medium be the message.

LETTERS, FONTS, AND STYLES

The fact that a computer program has options for many fonts, sizes, and styles does not mean that all are appropriate to use. (See the figure on the opposite page.)

Because it is so easy to generate attractive text, there is a tendency to pay less attention to brevity and clarity of word use. Hand-written text that is brief, well organized, and carefully planned is preferable to poorly planned, disorganized, overly wordy text that is beautifully printed.

It is this ease of text production that results in the many word slides shown in talks. For word slides, special care must be taken to keep statements short and to use simple and clear language as well as fonts. Labels for graphs and drawings should also be short, plain, and sized to be easily read.

USE COMPUTER OPTIONS TO SIMPLIFY AND ORGANIZE

For flow charts, tables, and diagrams, computer software offers almost limitless possibilities for rearrangement and variation of the composition. Take advantage of these assets to make the figure more easily understood. Use these assets to organize and simplify. (See the figure on page 110.)

DISTRACTING LABELS

ISOLATE OLIGOSACCHARIDES

DIRECT DETECTION OF LOW PICO-MOLE LEVELS OF CARBOHYDRATES.

SELECTIVE RESOLUTION OF POSITIONAL ISOMERS.

ANALYSE CARBOHYDRATES

High resolution ion exchange resins.

Pulsed amperometric detector.

SIMPLE LABELS

ISOLATE OLIGOSACCHARIDES

- Direct detection of low-picomole levels of carbohydrates.
- Selective resolution of positional isomers.

ANALYSE CARBOHYDRATES

- High resolution anion exchange resins.
- Pulsed amperometric detector.

On the left are some text options in the Microsoft Word® program. Upper case and mixed upper and lower case labels are available. The various styles (outline, shadow, underline, italic) say to the viewer, "Look how much the computer can do." Attention here is called to the unusual lettering, not to the information the text contains. On the right, the font is plain and sans serif; upper case is used for the two main headings. Information under the headings is upper and lower case and clarified by bullets and by being inset. Here the software was used to clarify and simplify information.

The computer is a flexible tool ready to do your bidding. But it will not plan, simplify, and emphasize information unless you tell it to.

COMPUTER DRAWINGS AND DIAGRAMS

Many computer programs can be used to make simple or complex drawings and diagrams. Among them are MacPaint®, MacDraw Pro®, CA Cricket Draw™, Canvas Draw™, Adobe Illustrator™, and Freehand® for the Macintosh. Good programs for MS-DOS® computers are CorelDRAW™, Arts and Letters®, MicroGrafx Designer®, and Adobe Illustrator™. Some programs, such as MacPaint® and MacDraw®, are easy to learn whereas others such as Adobe Illustrator™ and CorelDRAW™ are more complicated and difficult to learn.

However, use of a computer for a complex drawing may be time-consuming no matter which program is used. A hand-drawn figure may be quicker and just as effective. (See the figure on page 161.)

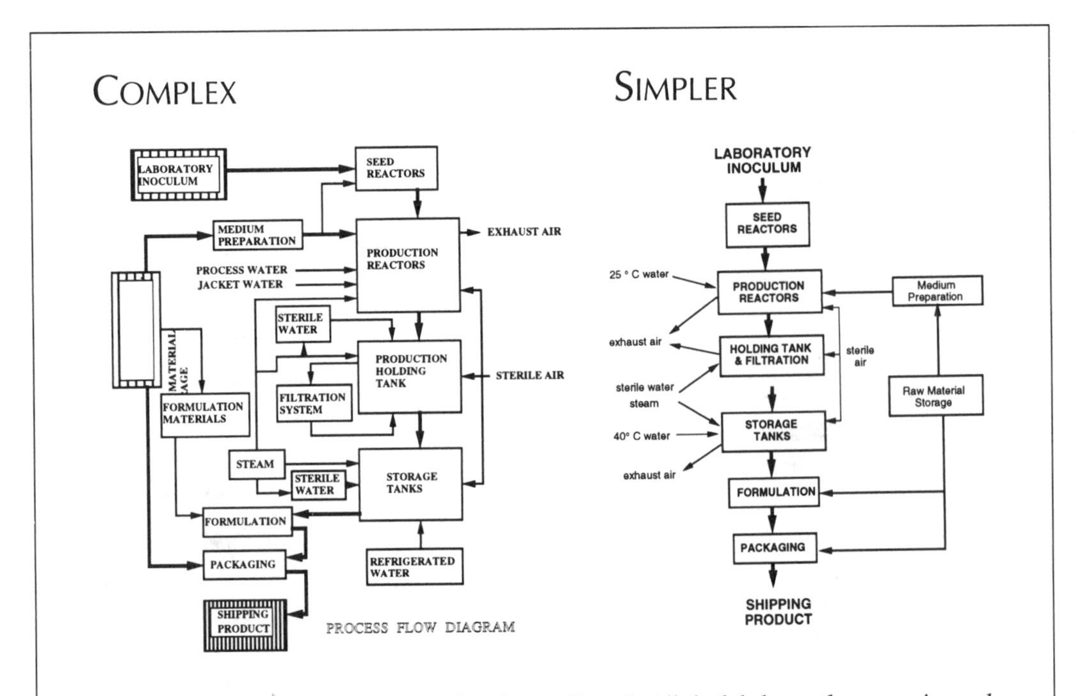

The flow chart on the left is cluttered and complicated. All the labels are the same size and style, and there is no focus for attention. To go from inoculum to product in this chart is to go through a baffling maze. On the right, the pathway from inoculum to product is straight and easy to follow. By the use of large bold labels, the emphasis is on this pathway. The plain upper and lower case labels to the right of the pathway are less emphatic, and the smaller unboxed labels to the left are least emphatic. Rearrangement of the flow into three columns of unequal importance has clarified the information and eliminated some repetition and clutter. Both charts were done in MacDraw Pro®.

The possibilities in drawing software are more than you will need. With imagination and experimentation, the results can be excellent and informative figures. The figures on pages 161–163 show some of these possibilities.

Not only does drawing software offer more options than needed for most projects, but the orientation of individual programs is often slanted toward advertising, publishing, or business. Generally, it is not necessary to learn every aspect of a program. Learning enough to do a specific project even on a complicated drawing program is not as difficult or time-consuming as might be expected.

Although drawings take time, it may be time well spent if it helps to clarify and focus ideas. Most of us are born artists, largely untrained. Doing drawings by pencil or computer may enable you to look at the information with fresh eyes. It can lead to a different way of thinking about the information and even suggest further paths to take.

HAND DRAWN

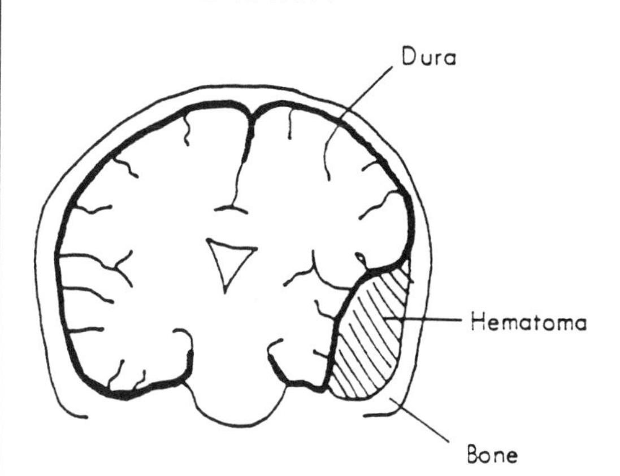

COMPUTER DRAWN

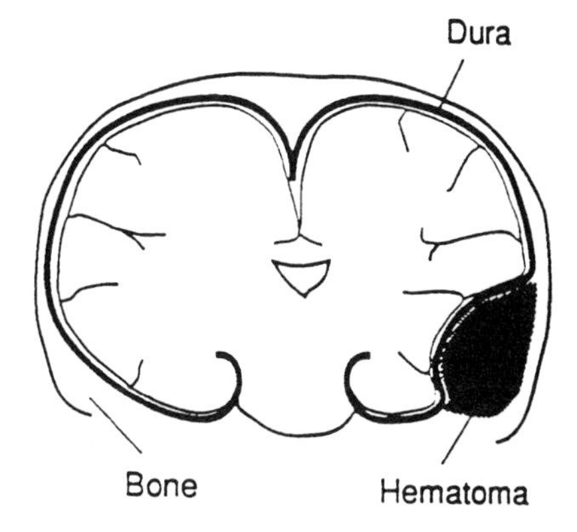

The drawing on the left was sketched in pencil, then overdrawn with a thick- and a thin-point felt-tipped pen. It was drawn to the size shown, and the labels were typed. It took about an hour. The computer drawing was done using the Adobe Illustrator™ program and took more than an hour and a half. Much more detail is possible in this program, but would add to the time involved.

ANATOMY

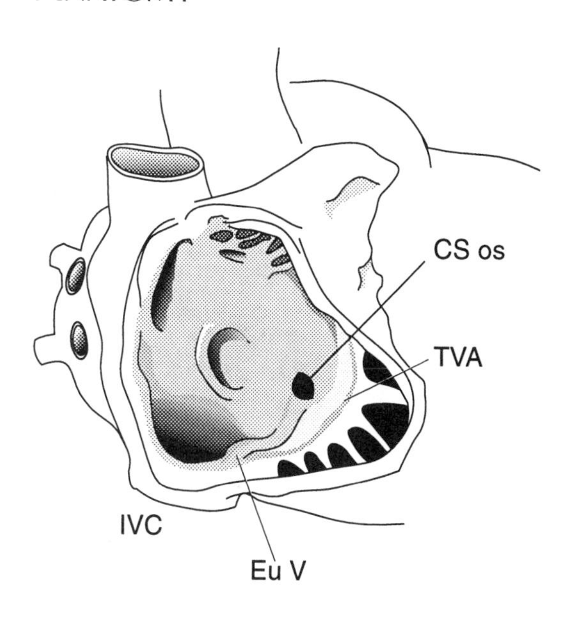

This preliminary drawing was done using Freehand® software. It underwent many changes in position, size, angle, shading, and lines before the final print. The time spent on this project was a way to learn the nuances of the anatomy as well as the software.

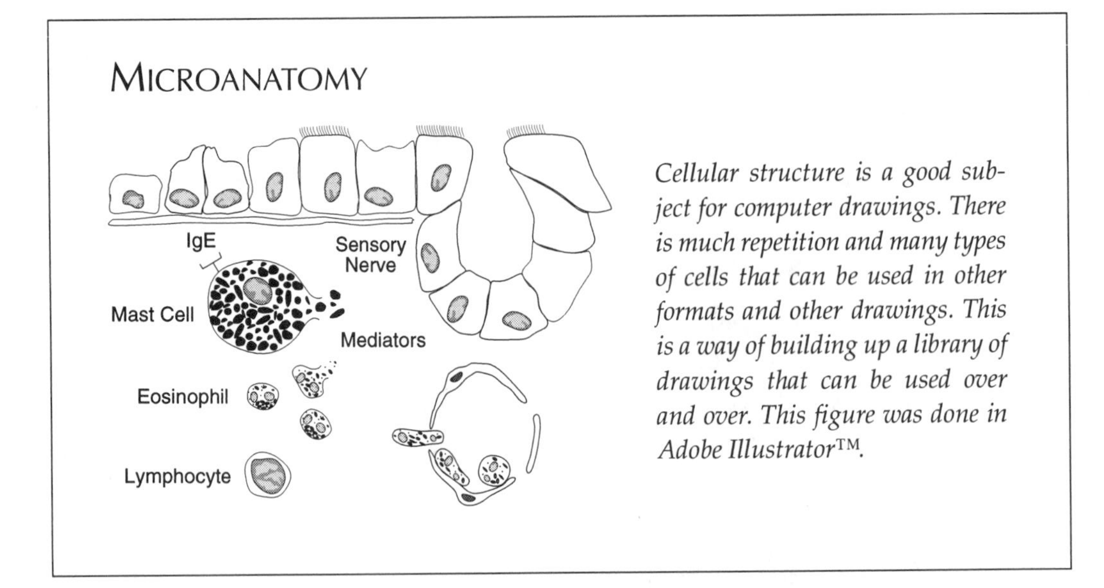

Cellular structure is a good subject for computer drawings. There is much repetition and many types of cells that can be used in other formats and other drawings. This is a way of building up a library of drawings that can be used over and over. This figure was done in Adobe Illustrator™.

The size and relative dimensions of computer drawings can be varied. They can be rotated, skewed, duplicated, shaded, and colored. They can be combined with other drawings or text. Elements can be eliminated. Drawings benefit from a good laser printer that produces even, crisp lines and will make many copies. Changes and additions can be quickly and easily reprinted.

If you do not need these advantages and do not expect to use the drawing again, a hand-drawn figure may be adequate and may save a little time. Also, some kinds of drawings are more suitable than others for the computer. Any drawing that requires strict accuracy or much repetition is much better done on the computer.

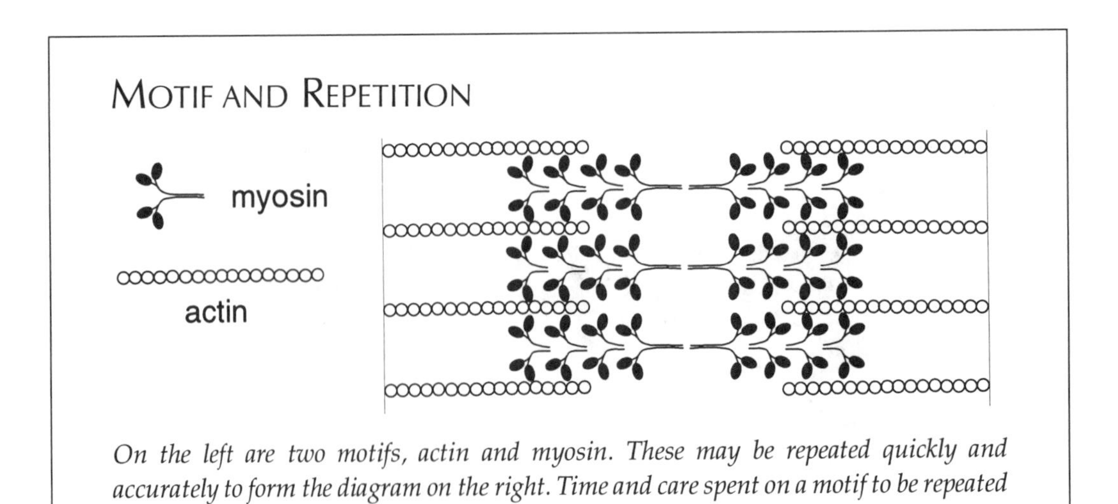

On the left are two motifs, actin and myosin. These may be repeated quickly and accurately to form the diagram on the right. Time and care spent on a motif to be repeated is well spent. The above figure was done on the Adobe Illustrator™ program.

A diagram that can be changed and extended is often valuable for communicating. This is easily done with a computer.

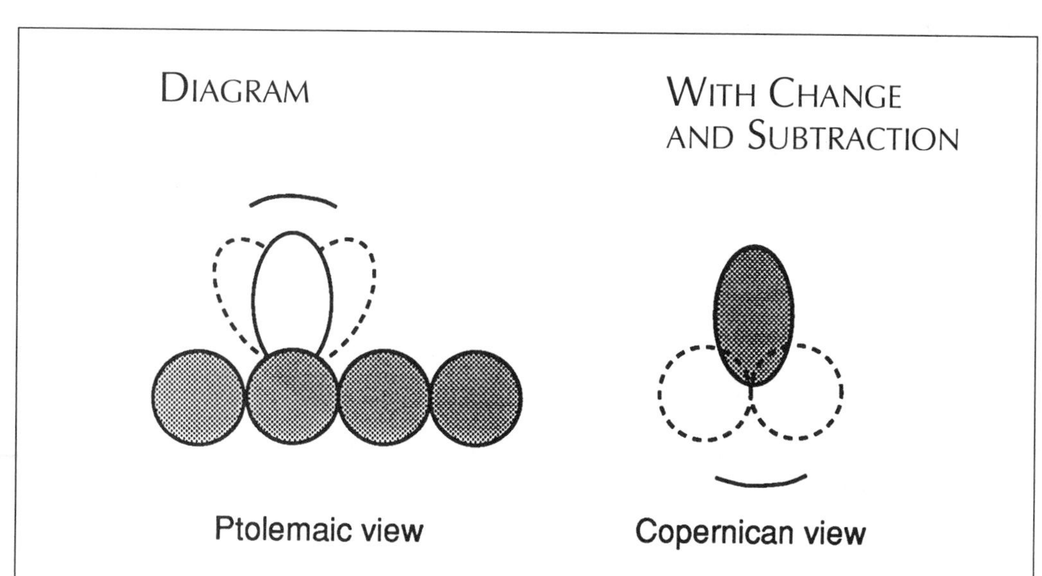

In the view on the left, the myosin moves while the actin is stationary, analogous to the sun moving around the earth. On the right, the myosin is stationary while the actin moves (earth moves around the sun). The basis of these diagrams is the ellipse and the circle. Lines, stippling, and positions have been changed and circles and ellipses added and subtracted. Both diagrams were done in the Cricket Draw™ program.

Most drawing programs offer options for exact rotating, reflecting, duplicating, and shading. These and other options such as variations of line thickness and style present numerous artistic and organizational choices. To make the right choices:

- Know exactly what you want to show.
- Try various ways of showing it. (The computer provides quick and easy ways to change.)
- Observe the differences carefully.
- If you are not sure about what works best, ask colleagues, artists, and friends for their opinion.

Scanners are becoming more affordable and allow copying of photographs, drawings, text, etc. The scanner software allows cropping, enlarging, change of contrast, and export to other software. Photodesign software such as Adobe Photoshop™ allows erasure, label addition, and additional drawing.

Scanners are useful for making slides from ultrasound, echocardiograms, X-rays, and gels. Once scanned, they can be cropped and labeled. Unwanted background and labels can be eliminated. Scanners are also helpful for customizing existing drawings and diagrams.

The drawing below was scanned from an existing diagram on an Apple One scanner into Ofoto software. It was then transferred into the Adobe Illustrator™ 5.0 program as a template.

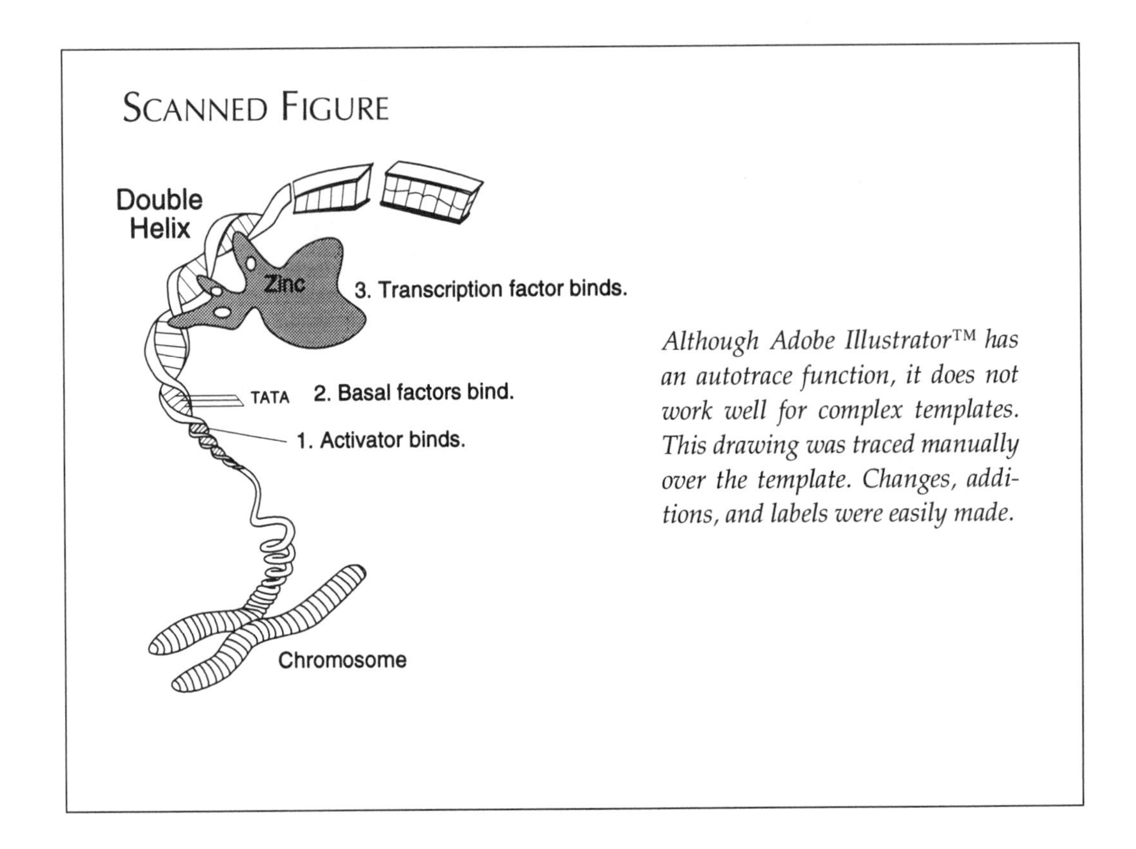

Although Adobe Illustrator™ has an autotrace function, it does not work well for complex templates. This drawing was traced manually over the template. Changes, additions, and labels were easily made.

One of the simpler programs that takes the place of the old MacDraw® is Claris Draw®. It is easy to learn and good for diagrams and simple drawings.

It can take time to learn to use the mouse for drawing. Mouse configurations differ, and comfort and ease of use are individual preferences. There are also affordable styluses and pads that more nearly resemble pencil or brush and might be easier for some to use.

Clip art, or ready-made drawings, is available and adaptable in some drawing programs such as Adobe IntelliDraw™ or Claris Draw®.

BIT MAPPED VERSUS VECTOR-BASED GRAPHICS

There are two kinds of computer-generated graphics, bit mapped and vector based. A discussion of the differences between the two may help decide which program to use.

A *bit-mapped* image is formed on the computer screen by a series of digital bits or square pixels. When printed, each of these bits corresponds to a printed pixel on the paper. MacPaint® or Paint Brush® are bit-mapped programs. Diagonal and curved lines in bit-mapped programs show noticeable stairsteps and appear jagged and somewhat crude. Enlargement of this kind of graphic exaggerates the "jaggies."

Although some designers use this jagged look to advantage, before deciding to use a bit-mapped program, decide whether it is appropriate for the subject.

Vector-based graphics (sometimes called line art or objects graphics) are mathematical descriptions of an object. These mathematical descriptions are represented on the screen by a series of dots. Output to the printed page is also a result of interpretation of the mathematical description to dots on the page. Diagonal and curved lines are correspondingly smoother, and enlargements do not suffer deterioration. MacDraw® and Adobe Illustrator™ are vector-based programs.

Drawing programs permit freehand drawing; graphing programs operate from data sheets to convert numbers to graphs. There is limited freehand drawing and annotation permitted in graphing programs.

COMPUTER-GENERATED GRAPHS

The computer is used widely in making graphs. A good graphing program offers flexible and speedy data input, a variety of kinds of graphs, and options for labels, axes, and symbol changes. Programs used for graph making are CA Cricket Graph™, DeltaGraph®, KaleidaGraph™, and Sigma Plot® for the Macintosh or Windows. Harvard Graphics® is designed for Windows. Spreadsheets, such as Lotus 1-2-3, Quattro, Excel, and StatView, have limited graphing capabilities. (See Chapter 6, "Graphs and Software," for more information on some of these programs.)

Input of data on the computer requires typing, proofing, and possible changes and rearranging of data. Plotting the graph requires the selection of axes and standard errors. In addition, changes are usually required in the axis lengths and units, label content, size, and style. It is possible to change the shape of the graph, save the format, or print it in a different size or alignment. Additions or errors in information can be corrected easily and replotted.

This takes time. Plan to spend an hour or more to make a publication-quality graph. In compensation, however, changes and additions may be made quickly.

Below are some steps in making a graph using the Cricket Graph™ program on the Macintosh computer.

DATA INPUT

	Minutes	Pre-resection	3 - 4 weeks	6 - 7 weeks
1	0.000	4.600	4.500	4.800
2	1.000	3.500	3.750	1.750
3	2.000	3.750	4.400	3.650
4	3.000	3.800	4.600	4.500
5	4.000	3.900	4.800	5.200
6	5.000	3.950	4.800	5.100
7	6.000	4.000	4.700	5.200
8	7.000	4.100	4.600	5.100
9	8.000	4.200	4.500	4.900
10	9.000	4.170	4.300	5.000

Data such as these can be imported from other software or it can be entered by hand. Corrections and additions are easily made. Often data are entered directly from the data recording equipment into a spreadsheet program. This is then transferred to a graphing program, saving time and ensuring accuracy.

UNEDITED GRAPH

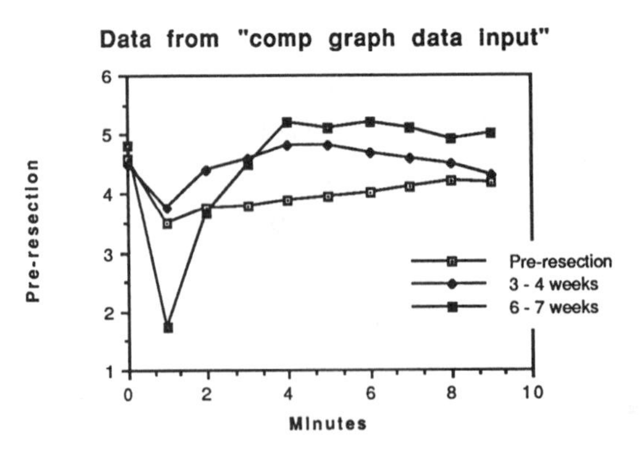

This is the graph showing the program default format. Notice that the axes are longer than necessary and have too many ticks. Labels are not accurate or complete, and symbols are not clearly distinct. There is an unnecessary plot frame, and the shape of the graph may be unsuitable.

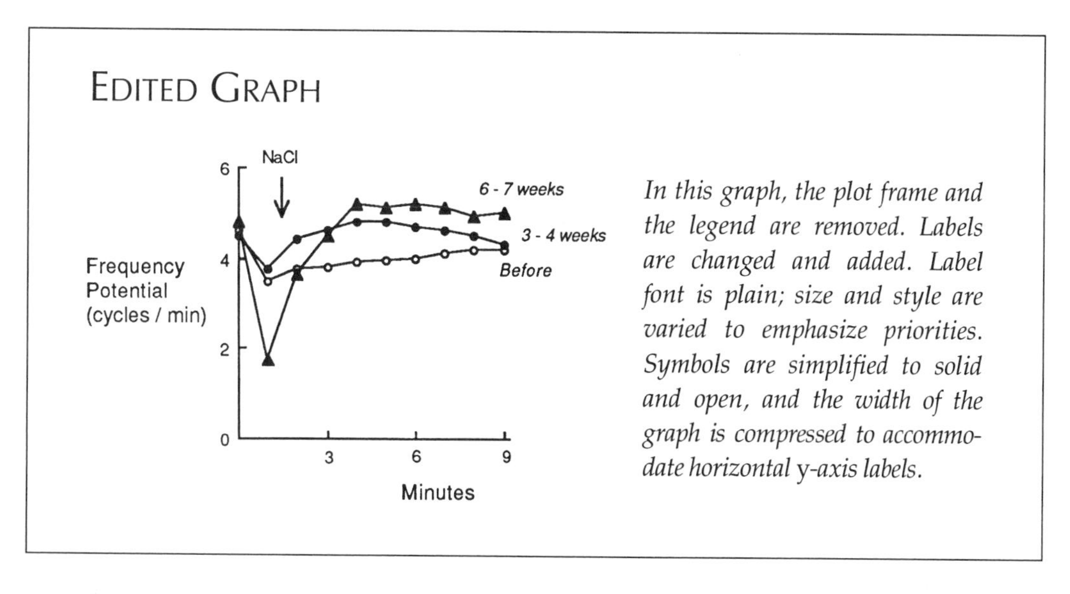

In this graph, the plot frame and the legend are removed. Labels are changed and added. Label font is plain; size and style are varied to emphasize priorities. Symbols are simplified to solid and open, and the width of the graph is compressed to accommodate horizontal y-axis labels.

Making changes that simplify, emphasize the important points, and are appropriate for the medium used is the most time-consuming but important part of making excellent graphs that communicate.

COMPUTER PRINTOUT

Both drawing and graphing programs permit output to *laser printers.* It is the laser printer that makes computer drawings and graphs viable for publication, slides, and posters. The quality of line and tone is good at about 300 to 600 dots per inch. Some of the inkjet printers give high, quality print and are excellent for figures that will be reduced. Enlargement shows fuzzy edges and lighter blacks.

PostScript is the language used to drive most laser printers, so that work done on the computer is converted by PostScript descriptions to fit the printed page. Although PostScript was developed originally as a typesetting language, it also permits output of vector-based graphics from both the Macintosh and MS-DOS® computers to the compatible laser printer. In addition, PostScript files also transfer to the typesetter's Linotronic printer. This printer generates 1200 to 2400 dots per inch and produces very fine quality print.

Use of good quality *paper* or film will affect the quality of the printed copy. The Linotronic printer uses a smooth, light-sensitive film, but it is possible to substitute special papers for the usual bond paper in the laser or inkjet printers. Some papers that help to produce dense black and crisply edged copy are

- Hammermill Laser Print—smooth, sharp image.

- Hammermill Laser Plus—one-sided bond with a slick surface, camera-ready quality.
- Hewlett Packard Plotter Paper—high gloss, heavy, for color and inkjet printers.
- Polaroid Overhead Transparency Film—clear acetate.

Conclusion and Suggested Readings

The multitude of choices facing you in making figures are both artistic and organizational. Whether you use a computer or another tool, it is important to know what you want to say, to try various ways to say it, and to observe what communicates effectively.

There are a wide variety of books on design and computer software. A few are mentioned below:

Harrison, D. Harvard Graphics 3: Self-Teaching Guide. Wiley, New York, 1992.

Jenner, D. Basic Tools & Techniques of Design for PC Software Arts and Letter Draw. Addison Wesley, Reading, Mass., 1994.

Lin, M.W. Drawing and Designing with Confidence. Van Nostrand Reinhold, New York, 1993.

McClelland, D. Photoshop 3 Bible. IDG Books, Foster City, Calif., 1994.

Paulson, E., et al. Using CorelDRAW. David P. Ewing, Indianapolis, Ind., 1994.

Perspection, Inc. & Microsoft. PowerPoint Step by Step. Microsoft Press, Redmond, Wash., 1994.

Swann, A. How to Understand and Use Design and Layout. North Light Books, Cincinnati, Ohio, 1994.

12 DRAWING BY HAND

Today, doing your own figures by hand seems old-fashioned. However, even now not every researcher has access to a computer or illustrator. So the ability to do your own figures by hand is a handy skill even for those with computers. It could even be an enjoyable skill.

Processes such as mounting and labeling gels and physiological tracings are not difficult to do. The process of making graphs becomes easy with practice, for some a relaxing change.

If you decide to do figures by hand, set aside time to do them well. If you have never done diagrams or graphs by hand, you need as much information as possible. Consult drafting and design books, chartists, artists, and people you know who have experience in doing their own figures by hand. You need patience and practice in addition to information.

Below are descriptions of useful and simple processes that anyone can do with a minimum of equipment and time. Tools and materials are also described. You may find that if you do not make your own figures frequently, tools and materials may be mislaid and your familiarity with them may have to be refreshed by practice before you begin. Take this into consideration when setting aside time to do your figures. Expect that you will make mistakes, so plan time for changes and additions.

TRIM, MOUNT, AND LABEL TRACINGS

Physiologic tracings are usually recorded on graph paper. Most of the time you will want to use only selected parts of the tracing to illustrate your point. You may want to eliminate single lines or whole sections of multiple tracings. You may want to combine scattered parts of the tracing. Also, for clarity of communication, tracings should have scales and brief labels. Even if a computer directly prints out physiological recordings, it may be easier to combine, eliminate, and label parts by hand.

Tools and materials* that you will need for this are

- A T-square, preferably metal.
- A drawing board against which a T-square will fit and on which tracings may be cut.
- A triangle, preferably metal.
- A cutting knife (e.g., #1 X-acto knife) with #11 blades.
- Two-ply, plate-finish bristol paper.
- Masking tape.
- A technical pen, (e.g., Rotring or Koh-I-Noor Rapidograph) in a fine-point size. #0 (0.35 mm) or #1 (0.50 mm) are good sizes.
- Black drawing ink (e.g., Pelikan) and liquid pen cleaner (e.g., Rapido-eze or Speedball).
- Sheets of transfer lettering (e.g., Letraset or Chart Pak).
- Spray adhesive (e.g., 3M Photo Mount).

Before beginning, use a photocopy machine to make practice copies. (See the figure on the opposite page.) A photocopy machine is also valuable for making the fine lines of the tracing darker and thicker so that you may want to mount the copy rather than the original. A photocopy machine will also drop out light, colored grid lines, and some machines will reduce or enlarge the tracings.

Trim the Tracing (See Figure on Page 172)

- Align the tracing with the T-square and tape it to the board.
- Using the cutting knife and a sharp blade, make horizontal cuts along the T-square.

*Materials listed are available in art and drafting supply stores and in some stationery stores. Also various catalogs are available for mail orders. For a copy of the Flax Artist Materials catalog, call (800) 343-3529.

PHOTOCOPY FOR PRACTICE

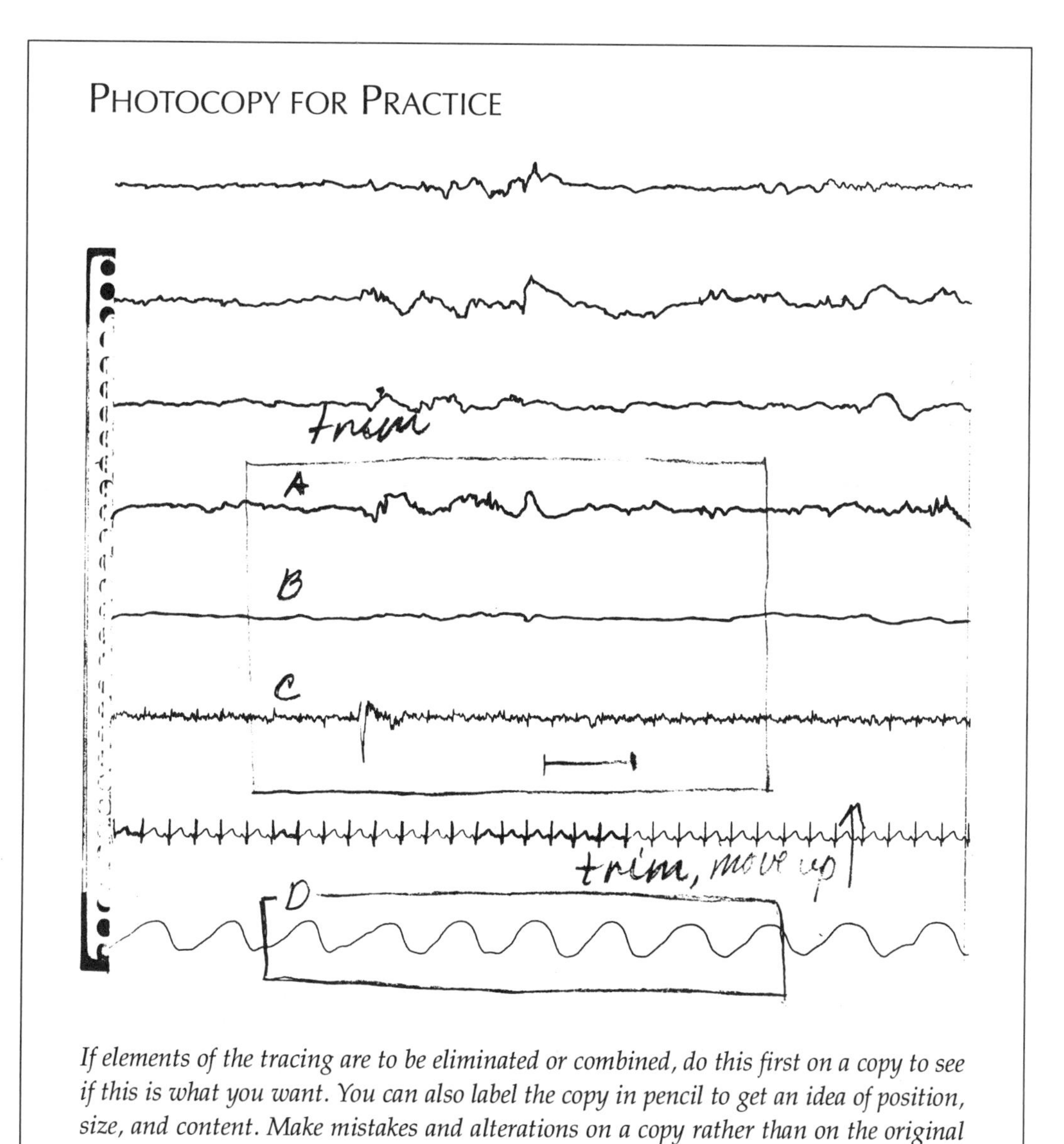

If elements of the tracing are to be eliminated or combined, do this first on a copy to see if this is what you want. You can also label the copy in pencil to get an idea of position, size, and content. Make mistakes and alterations on a copy rather than on the original tracing.

- Place the triangle against the T-square for a perfect right angle. Make vertical cuts along the triangle.

Mount the Tracing

- Tape the bristol paper to the board.
- Loosely position the cut tracings as desired on the board and mark their final positions lightly with pencil.

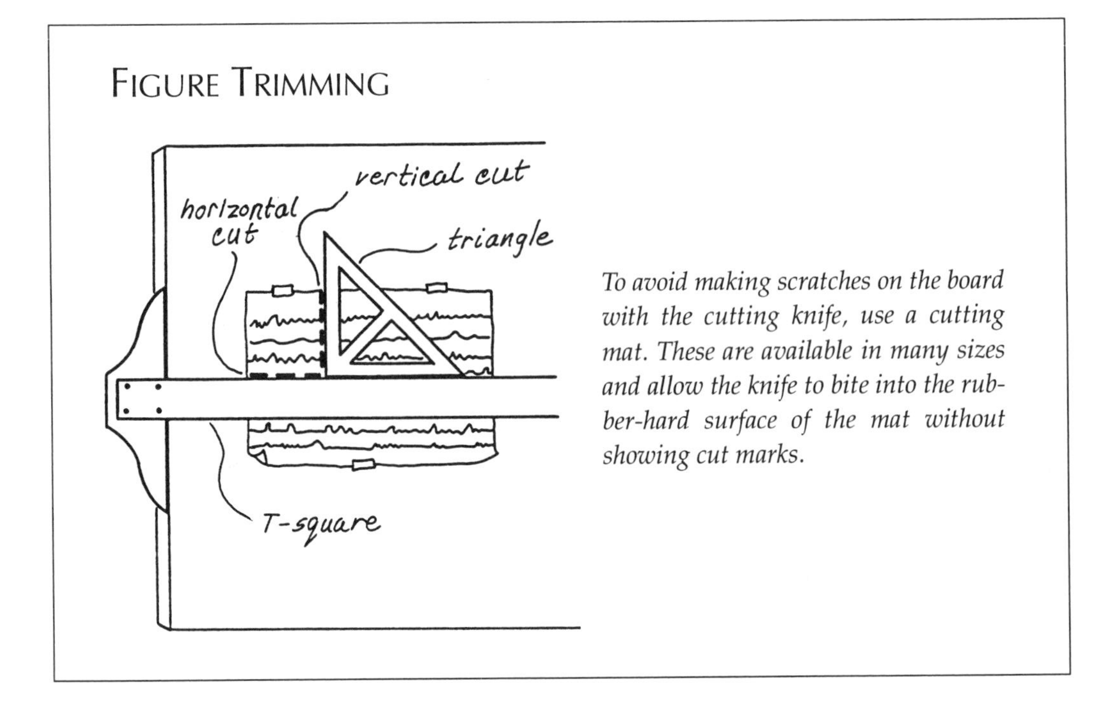

FIGURE TRIMMING

To avoid making scratches on the board with the cutting knife, use a cutting mat. These are available in many sizes and allow the knife to bite into the rubber-hard surface of the mat without showing cut marks.

- Using the T-square and triangle, pencil in light vertical and horizontal lines indicating the desired positions of the trimmed tracings.
- Lay the tracings face down on newspapers that have been spread out to protect floor or desk surfaces from excess spray adhesive. Spray the backs of the tracings with spray adhesive following directions on the can.
- Position the sprayed tracings along the lines marked on the bristol paper. If a tracing is misaligned, lift it carefully from the paper and reposition.
- After tracings are in position, cover them with clean paper and rub them down with a thumbnail or a burnisher. (3M makes a 4-inch-wide burnisher that is excellent for this purpose.)
- Erase all pencil lines. If there is adhesive on or around the tracings, it may be removed with a gum eraser or small amounts of rubber cement thinner.

Mark Units

Because the lines of tracings are fine, their unit markings and labels should also be fine so that they will not compete with or overwhelm the tracings themselves. Use a fine-tipped technical pen (#0 or #1 point) for scale markings, axis lines, and tick marks. This kind of pen must be held perpendicularly to the paper between thumb and forefinger. The pen should be angled away from the T-square and triangle to prevent ink from running under them. Or put layers of masking tape on the bottom sides of the T-square and triangle to lift them from the paper. Take the time to practice drawing straight lines against the T-square and triangle.

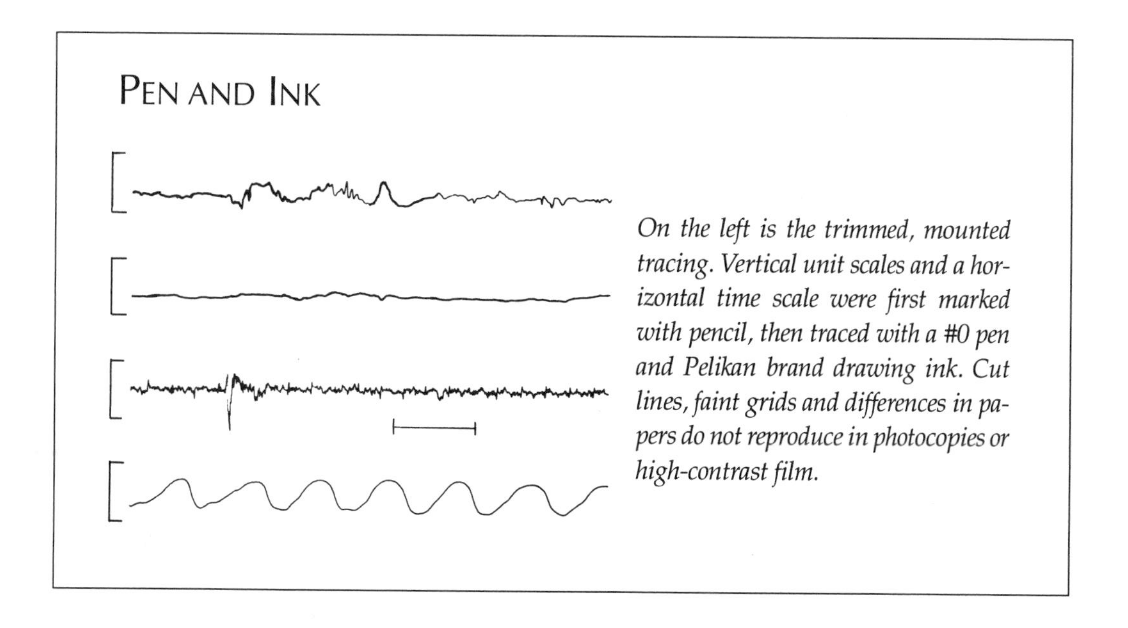

On the left is the trimmed, mounted tracing. Vertical unit scales and a horizontal time scale were first marked with pencil, then traced with a #0 pen and Pelikan brand drawing ink. Cut lines, faint grids and differences in papers do not reproduce in photocopies or high-contrast film.

Pens have to be kept clean and must be handled carefully. If you prefer not to use pen and ink, other options are

- Pressure sensitive graphic *tape* in a #1 point size (0.35 mm). This may be positioned on your marked pencil line and cut to the appropriate length.
- *Transfer lines* (e.g. Letraset) are available in various thicknesses (#83-106), or 0.5 point only (#4269), or 1 point (#4271). These lines may be cut through the plastic sheet in any length and rubbed onto the penciled lines with the point of a pencil or a special burnisher (sometimes referred to as a spatula) for this purpose. Use nonphoto blue Berol Prismacolor or Berol Verithin pencil lines for marking position lines since they do not need to be erased. Nonphoto blue is a particular shade of blue which does not reproduce on film. Erasure of lines near transfer labels risks removing the labels.

Label the Tracing

Transfer type comes in various sizes, fonts, and styles. Helvetica Light or Helvetica Medium is a sans serif font that is readable and attractive. A light or fine-lined font works best for fine-lined tracings. For an average 8.5 × 11-inch tracing, 18-point size works well. However, it is wise to buy several sizes: 18 point for the main labels, 14 point for numbers, 24 point for titles. Stick to one font for consistency and make sure you will have enough of the required letters and numbers.

Before applying the labels, use the T-square to draw a light blue pencil line slightly below the label position. There is a dashed line below the letters on the sheet. Align this dashed line along the pencil line while you position the first letter. With your finger, press the letter into contact with the paper. Then rub back and forth across the letter with the pencil point or spatula using light pressure. Peel back the sheet and the letter is transferred. Do the same for the subsequent letters until the word is complete. When the word or sentence is finished, cover it with the blue backing paper and burnish or rub it with the spatula or your thumbnail.

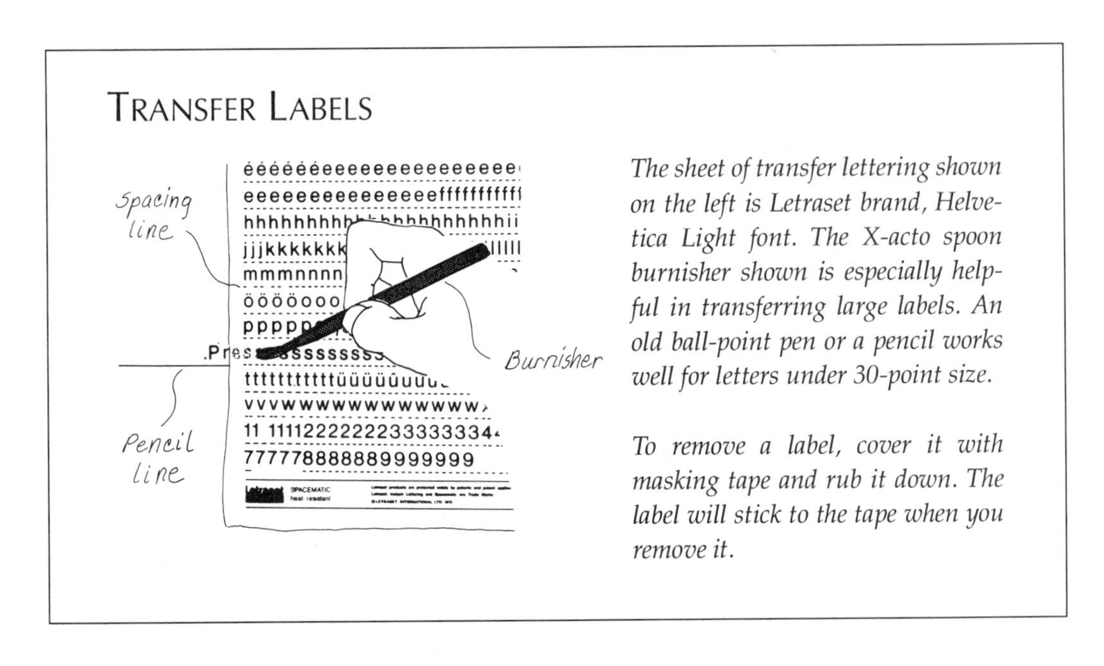

TRANSFER LABELS

The sheet of transfer lettering shown on the left is Letraset brand, Helvetica Light font. The X-acto spoon burnisher shown is especially helpful in transferring large labels. An old ball-point pen or a pencil works well for letters under 30-point size.

To remove a label, cover it with masking tape and rub it down. The label will stick to the tape when you remove it.

The dashed lines below the letters are actually marks for spacing. If you need help spacing the letters in a word, transfer the right-hand space bar (dash) under the letter along with the first letter. Abut the left-hand space bar of the next letter against the previous space bar. Transfer this letter and its right-hand space bar and do the same for the following letters of the word.

If rubbing does not transfer the letters or if the transferred letters appear cracked or uneven, it may be because the sheet is old. You can try lightly rubbing the letter on the blue backing sheet before transferring it, and holding the Letraset sheet in position while carefully peeling it back. Or to spare yourself frustration, throw away the old sheet and buy a new one.

You may find that you use up certain letters or numbers quickly, leaving the rest of the sheet unused. This makes Letraset an expensive form of labeling (the cost is now about $14 per sheet). It is also time-consuming, especially if labels must be centered. To center accurately, roughly trace the letters of the word or sentence in pencil on a tissue overlay. Position the traced word or sentence on your paper and mark the beginning position for transfer.

Other possibilities for labeling are the lettering systems such as *Leroy* or the *Kroy* labeling machine. The Leroy system uses a scriber and various size templates to ink directly on the paper. The Kroy machine prints labels on adhesive-backed tape, which is then cut and positioned on the paper. (See the figure on page 176.)

LEROY SYSTEM

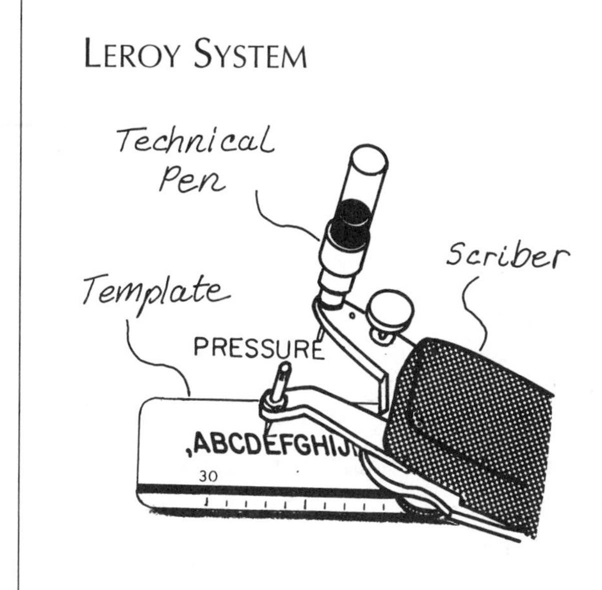

With practice, this system becomes a versatile, quick, and inexpensive way to label. Although the font is unchangeable, the style may be changed to italics, and pens may be interchanged to produce thicker or thinner lines.

The problem with centering is the same as for transfer type, and mistakes must be erased or covered with opaque white-out.

Although centering is easy in the Kroy system, the edges of the tape will show in continuous-tone photographic reproduction. Both the Leroy and the Kroy systems have a high initial cost, which can be justified if you do frequent labeling.

Computer-generated labels printed on a laser printer look professional and may be printed in all the usual sizes and fonts. The back of the labeled sheets can be spray-glued and cut out, positioned, and rubbed onto the figure. Or use a glue stick on the cut-out label and affix.

For frequent gluing of small items such as labels, a hand-held waxer is useful. It heats wax electrically and rolls out a thin coat onto paper. Once the waxed item is pressed into position, it stays well but is also easy to peel off and reapply if changes are required.

KROY LETTERING MACHINE

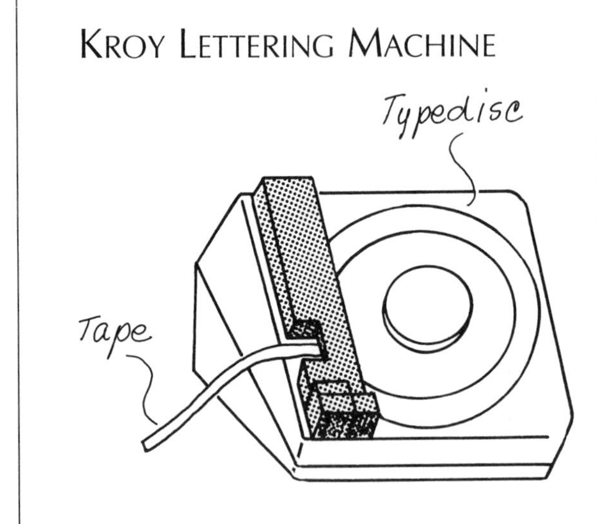

The Kroy machine might be compared with a typewriter. Characters are automatically spaced and aligned. A carbon impression of each character is pressure fused to an adhesive-backed tape. The tape backing paper must be removed before it is applied to the desired position. It can be repositioned and removed easily. Several kinds of tape are available, including color tape. Each typedisc represents one style in one size.

TRIM, MOUNT, AND LABEL GELS

The process described above for tracings may be used also for photographs of gels. It is a good idea to photocopy the gels and trim and label the photocopy to experiment with spacing and labels.

Because photographs of gels are generally no larger than 5 × 7 inches, labels should be small. Use Letraset point sizes 10, 12, and 14. If you use more numbers than letters, you might experiment with sheets of numbers only. Numbers in Helvetica font come in various point sizes but in medium thickness only. Futura Medium is a lighter type face than Helvetica yet is compatible with it.

Indicator lines may be drawn with a fine-point technical pen or rubbed on from sheets of transfer lines. Remember to leave space around the trimmed gel photos so that two negatives may be made of the final gel. (See "Chapter 3, Photographs.")

The labeling of gels has been a lengthy process, and changing labels is tedious. If a scanner is available, it is possible to scan the gel at 100% and use the resulting scanned image as a template for labeling. The photograph of the gel can be then pasted over the scanned gel image. Changes require that the photo be removed and reglued to the changed template.

DRAW AND LABEL GRAPHS

Sometimes it is quicker to draw a graph by hand than by computer, especially if you have already plotted it on graph paper. Detached axes, composite graphs, special curve fits, and graphs with special axes all usually require transfer to another drawing program for additional time-consuming changes.

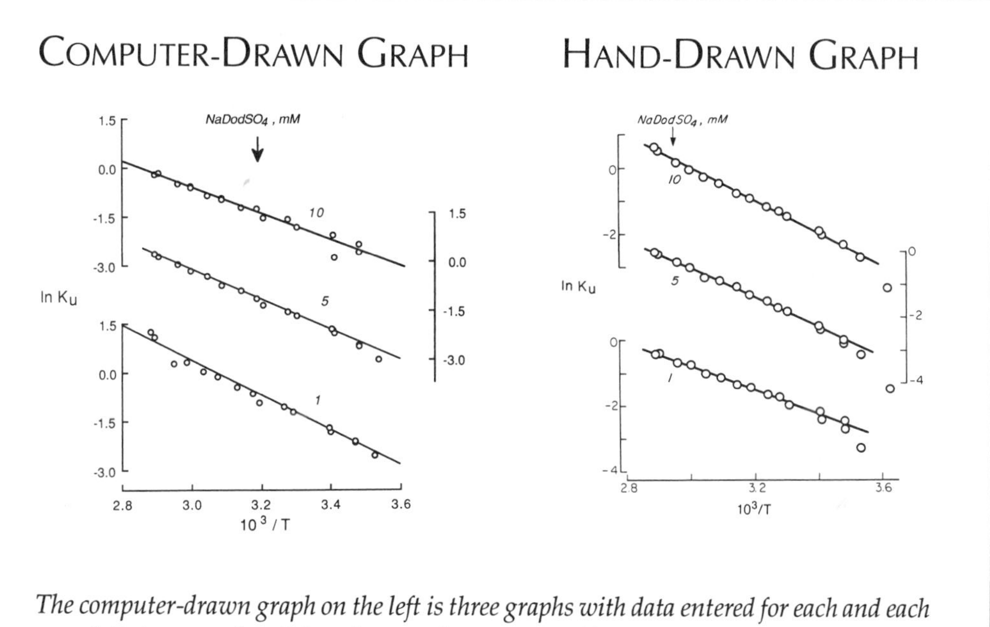

The computer-drawn graph on the left is three graphs with data entered for each and each one plotted separately. Although some time was saved by using the same format for each graph, all three had to be transferred to a drawing program. Total time to do this is close to 2 hours. The similar hand-drawn graph on the right can be done in less time if you have had some practice drawing graphs.

If changes are required, time saved in the hand-drawn graph will be lost, and the erasures, pastes, and cuts required for the usual changes may show up in photography.

Sometimes it is satisfying to spend time working directly on paper, using hand tools. Drawing graphs is a relaxing change for some people that may even be therapeutic. However, graphing programs now offer so many options for change that computer-drawn graphs can be as aesthetically pleasing as a good hand-drawn graph. Whatever method is used, a good graph should be simple and uncluttered, focused on the main points, and designed for the medium used.

With practice and proper materials and tools, graphs may be made quickly by hand. Many of the *tools and materials* listed above for mounting and labeling tracings may be used also to make graphs. In addition to the board, T-square, triangle, paper, and masking tape, you will need the following:

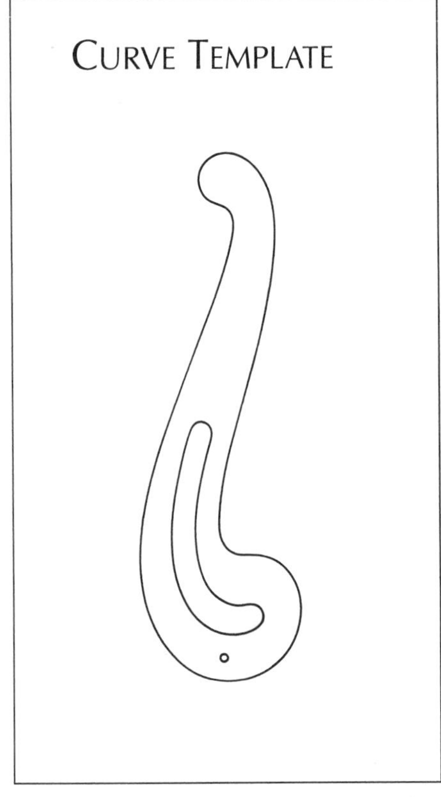

- A "light box" suitable for use with the T-square.
- Or transparent film or vellum for copying without a light box.
- Several technical pens of different point sizes (i.e., #1, #2, and #3 point pens).
- A symbol template (e.g., Sketch Mate Rapidesign, #19).
- Or a sheet of transfer symbols (e.g., Letraset key symbols, #2453).
- A French curve in a shape similar to the one shown to the left.
- Or pressure-sensitive, flexible black tape in 1⁄16- or 1⁄32-inch widths (e.g., Flex-a-Tape).
- An electric eraser and/or white paint (e.g., Pro White opaque watercolor) and small brushes (e.g., Winsor Newton or Grumbacher red sable brushes, sizes 0, 00, or 000) for corrections.

Data should be plotted on graph paper in pencil, then copied in ink onto good paper. If you have a light box, use a plate-finish bristol paper, one or two ply. Tape the bristol paper over the graph paper on the light box. When the light is turned on, it is easy to see the penciled graph lines for copying in ink.

Polyester drafting films such as frosted Mylar or a good-quality, transparent vellum such as Vidalon can be used without a light box. Simply tape the film to the board over the penciled graph and copy in ink.

Axis Lines and Ticks

For axis lines and ticks, use the #1 and #0 pen, respectively. Draw slow and steady lines against the T-square and triangle in the same way described on pages 172 and 173 for tracings.

Design Symbols

For symbols, use the symbol template with a #2 pen. Practice with it first. You may find that if you fix tape to the solid parts of the template, it will lift it off the paper enough so that ink will not seep under the template. The best symbols to use are

- Open and closed circles. Solid circles seem larger than open circles. For equal emphasis, make the solid circle one size smaller.
- Open and closed triangles. Triangles in general look smaller than circles, and appropriate adjustments should therefore be made.
- Open and closed squares. Squares look largest of all the symbols.

In reduction, it may be difficult to distinguish between circles and squares. Triangles and circles are the easiest to distinguish in reduction.

SYMBOL TEMPLATE

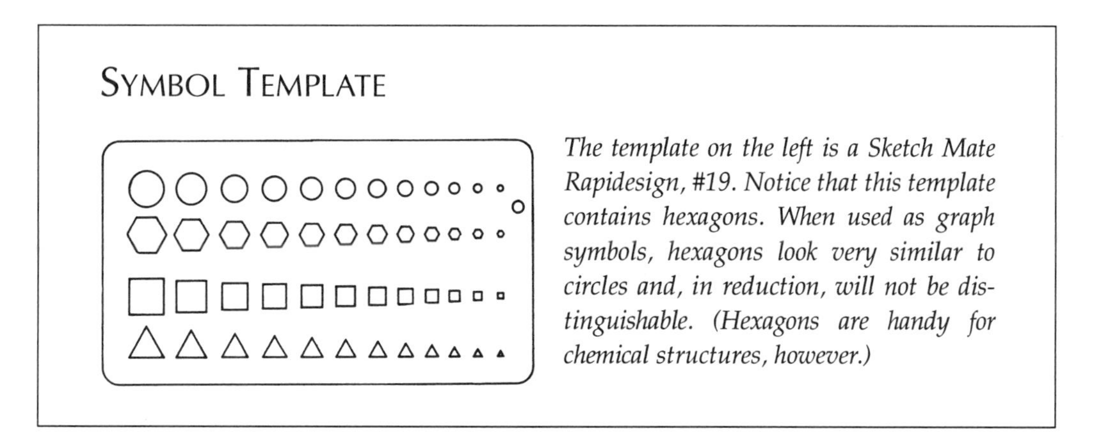

The template on the left is a Sketch Mate Rapidesign, #19. Notice that this template contains hexagons. When used as graph symbols, hexagons look very similar to circles and, in reduction, will not be distinguishable. (Hexagons are handy for chemical structures, however.)

Straight Lines

Straight lines connecting points are easily drawn with a ruler and a #3 pen. If you stop short of the closed symbols, the symbol becomes better defined.

SPACE AROUND CLOSED SYMBOLS

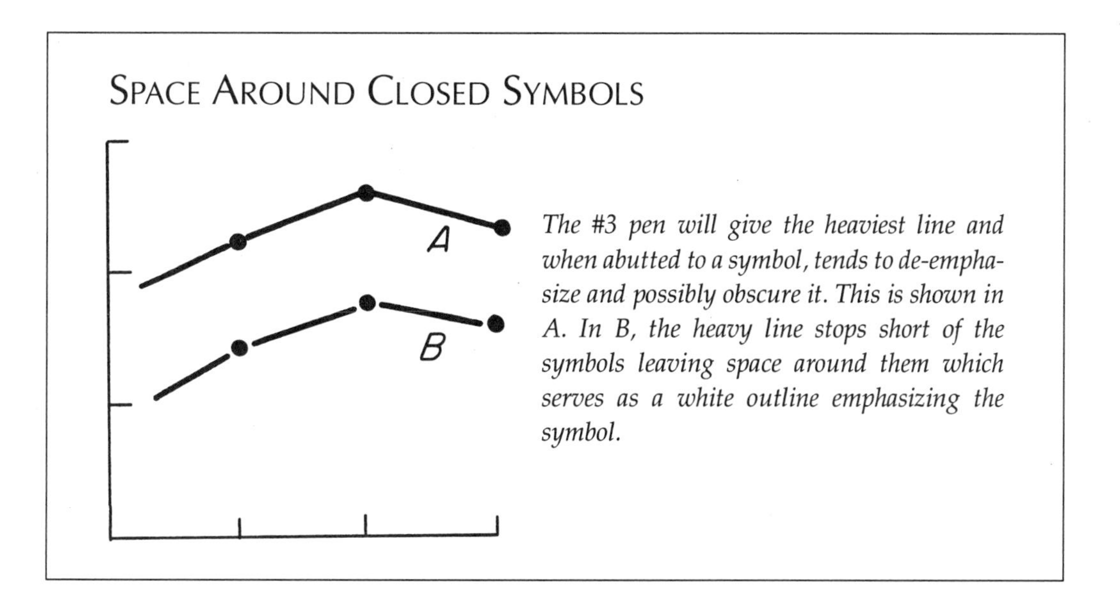

The #3 pen will give the heaviest line and when abutted to a symbol, tends to de-emphasize and possibly obscure it. This is shown in A. In B, the heavy line stops short of the symbols leaving space around them which serves as a white outline emphasizing the symbol.

Curved Lines

Curved lines done with the #3 pen against a *French curve* take practice. The French curve will generally only fit short segments of the curve, so that the curve will have to be repositioned often. It takes practice to draw the continuation of a line without being slightly offset from the original line.

To make a smooth curve quickly, use *flexible tape.* Press the tape to the beginning of the curve, leaving some excess, which may be trimmed exactly later on. Do not stretch the tape by pulling. Gently lay the tape along the curve, pressing down with your finger when it is in the proper position. Trim the beginning and end of it with the cutting knife and a sharp blade. The tape must be patiently eased, and sometimes even cut, to accommodate sharp curves.

Label the Graph

Label the graph with any of the methods described on pages 173–175 for tracings. The numbers and labels on the face of the graph, as for curves, should be the smallest size. The axis labels should be larger and are generally easier to read if they are upper and lower case. Make the title all upper case so that it will be the largest and most prominent label on the graph. (Or use upper and lower case in a larger font size.) It is preferable not to use bold labels because they compete with the graphed data on the face of the graph. They also tend to thicken and blur in reduction.

Corrections

Corrections are almost always necessary, and there are many methods for changing and improving.

- Small bumps or blobs on lines may be cleaned up with a small brush and white opaque paint. Use a magnifying glass for fine corrections.
- Irregular thicknesses of straight lines may also be covered by applying transfer white lines such as Letraset #223 arrows and lines in white. Cut the line to required dimensions and position it next to the straight line, covering the irregularities. Gently rub the line to transfer it.
- A symbol may be changed by applying white correction fluid (Lindsay's is one brand) and drawing another symbol over it after it has dried. Correction fluid, when dry, has a smoother surface than watercolor paint for accepting ink.

- A transfer label can be removed by covering it with masking tape and rubbing it down. When the tape is pulled off, the label will come with it.
- Lines, labels, symbols, and smudges may be removed by an electric eraser. With light pressure, it will remove ink from bristol, film, and vellum. To remove small areas of ink that are next to other areas you want to keep, use a metal eraser shield with the electric eraser.

USE A PHOTOCOPY MACHINE FOR CORRECTIONS AND CHANGES

Many photocopying machines produce copies with excellent line quality that are suitable for print, slide, or poster. Many photographers prefer this kind of copy to originals with taped edges, white-outs, cut-and-pasted areas, and thin or gray lines. For slides, color may be added to photocopies using color film or markers. Photocopied transparencies may be used for overhead projectors or for figure overlays. Many photocopy machines permit enlargement or reduction, and enlarged photocopies may be used for posters. Reduced photocopies are helpful for layouts, projections of reduction, insets, or making labeling changes.

A photocopying machine can be very helpful to

- Eliminate unwanted material. Tape white paper over the unwanted area or cut it out and mount the desired area on white paper. Photocopy it using a normal setting.
- Add or exchange material. Cut and paste or tape additional or substitute material onto an existing figure. Photocopy the changed figure.
- Move material around and combine material by cutting and pasting.
- Enlarge or reduce labels by pasting or taping new labels over the old ones. This is illustrated on the next page.

It is easier to cut and paste new labels than to redo them.

There are many ways to use the photocopy machine creatively to improve figures. For instance you can make a journal figure into a creditable slide by covering the caption and cutting and pasting or typing new labels and adding a title. (See Chapter 8, "Slides," pages 126–127.)

With practice, you will develop skill and find shortcuts and new materials that will help you to do figures quickly and professionally. Use your imagination and visit stationery and art supply stores and study catalogs for new materials and tools that will serve your purpose.

LABELS TOO SMALL IN REDUCED GRAPH

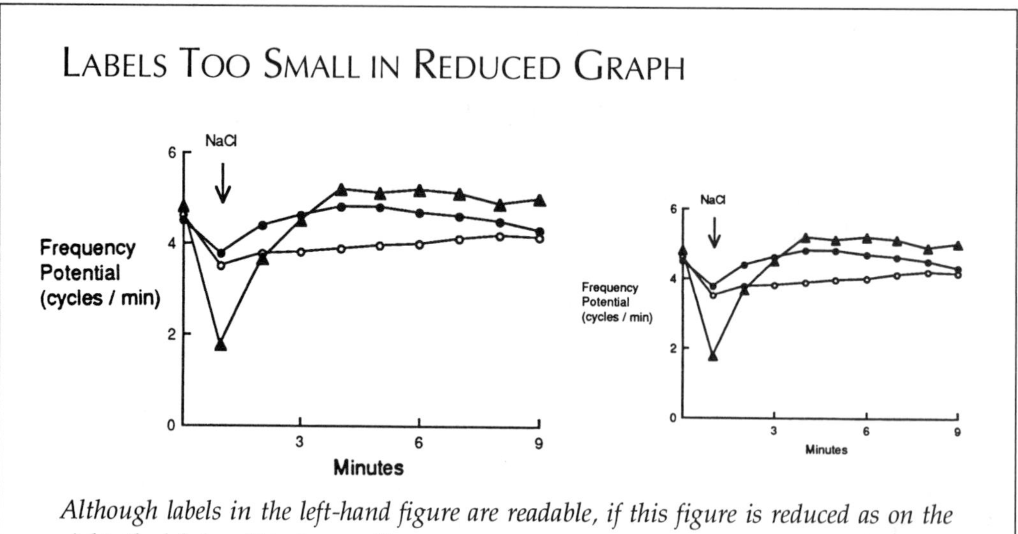

Although labels in the left-hand figure are readable, if this figure is reduced as on the right, the labels will be too small.

NEW LABEL ON REDUCED GRAPH

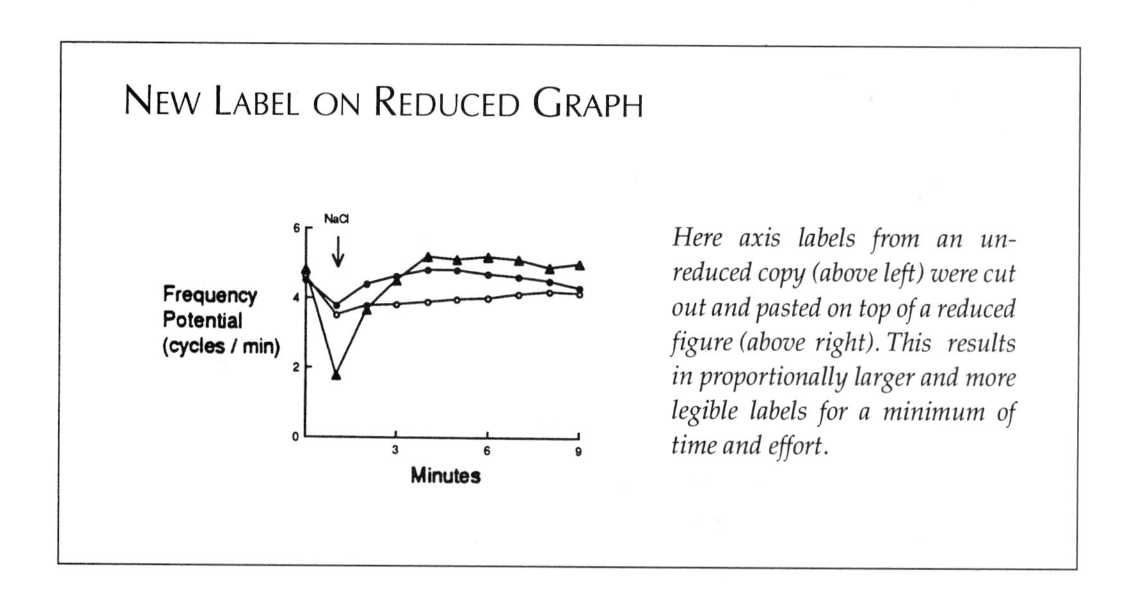

Here axis labels from an unreduced copy (above left) were cut out and pasted on top of a reduced figure (above right). This results in proportionally larger and more legible labels for a minimum of time and effort.

Skill in the use of tools and materials for drawing give you independence and flexibility in getting figures done. Doing your own figures will also increase your knowledge and appreciation of good illustrations and make you more visually discriminating.

13 Conclusion

How often have I heard a researcher blame the audience for its failure to understand: "The information is obvious. If they don't understand it, that's their problem." In reality, the "problem" is the communicator's, and it is serious.

Communicating information is not the same as doing research. Your data may be excellent and your analysis superb. You may be creative and imaginative at doing all the things necessary for fine research. No matter; these valuable traits will not automatically translate into clear communication.

It is ironic that today the best communicators communicate the most trivial information. Advertisers and ad agencies depend for their livelihood on selling an idea or product. They spend great amounts of time, energy, talent, and resources to do this. (See the top figure on page 184.)

By contrast, many scientists spend time, effort, and money to obscure and complicate serious and sometimes vital information. (See the bottom figure on page 184.)

Advertisers do not try to impress with learning. Advertisers do not try to confuse. Although some heap scorn on advertisers, we can learn a great deal from them about communicating.

You Should Learn to Communicate

In his book *Information Anxiety,* Richard Saul Wurman talks about how we are "inundated with facts but starved for understanding." This sort of starvation is not

ADVERTISEMENT

It's sort of like a tooth vitamin.

Crest similar to a vitamin?

Well, vitamins help keep your body strong. Crest does the same for your teeth. Vitamins help keep your body healthy. So does Crest for your teeth. Vitamins build up your body's resistance to disease. Crest builds up your teeth's resistance to cavities.

What makes Crest special is its fluoride, Fluoristan.* And yet, most toothpastes don't have it. And that includes the leading toothpastes.

For example, take the five leading toothpastes. Only one contains the "tooth vitamin."

Of course, you know which one.

Ingenuity as well as thought, time, skill, and money went into this ad. The layout is simple and attractive; the analogy of toothpaste to a vitamin is boldly and provocatively presented; the photograph reinforces that analogy; the text is concise. This figure attracts attention, then follows through with a clear, simple message.

SCIENTIFIC PRESENTATION

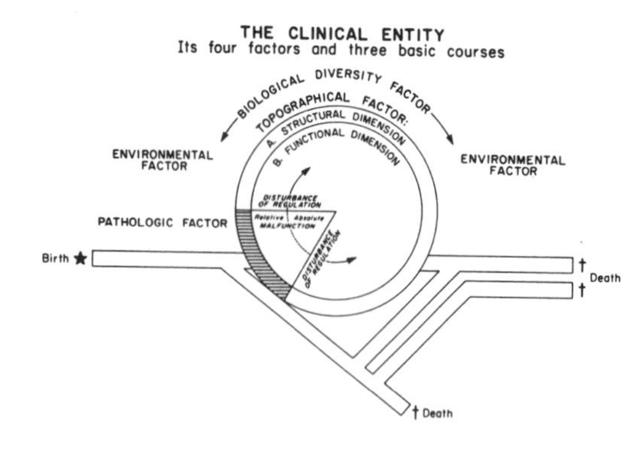

This figure is pretentious. Its simple message is that the course of disease has four factors and three outcomes. The wording obfuscates and the symbolism misleads.

only sad but also dangerous for scientists. As the gap between the scientist and the public grows, public anxiety about scientific research increases and groups such as the animal rights enthusiasts gain strength.

This sort of starvation is also apparent in the failure of scientists to communicate clearly with other scientists. There is increasing specialization in scientific research and much greater complication of information. Unless a scientist is very clear and simple in the presentation of information, even other scientists may not understand.

This matters. It matters because the sharing of information among scientists is important for scientific progress. It matters because grants and jobs in science depend on communication. It matters because understanding of the world in which we live is important for our survival.

YOU CAN LEARN

We recognize a special skill involved in writing and speaking clearly and simply. There is also a special skill in presenting visual information clearly and simply. Fortunately, it is a skill that can be learned. It is a skill that the best of scientists have used and thought about, and one that is extremely useful for the furtherance of science, career, and general knowledge. The drawings of Vesalius and his artists still speak to us, although his words do not. Pictures by Audubon and pictures of molecular structure are illuminating and astonishing.

Awareness

Awareness of standards, requirements, and means of making illustrations is the first step in learning visual communication.

- To use the services of an artist or photographer, you must be aware of what makes a good illustration.
- You must be aware of the limitations and assets of the media and materials that will be used.
- You must be aware of the illustrator's and photographer's informational needs and methods of working.

Awareness of your audience's background and level of knowledge will determine the kind of illustration that will communicate effectively.

Awareness of available choices is essential for good communication. This knowledge of existing choices should lead in turn to experimentation with new ways to express information visually.

Clear Purpose

If your own purpose is not clear, your figures will be confusing. Do not be afraid to be simple.

- Your general purpose should be to convey understanding through pictorial means.
- The purpose of individual figures must be clear and well thought out.
- Limit the amount of information in each figure.
- Emphasize what is important.

Time, Patience, and Practice

- Time to observe and think about your figures.
- Time to plan and organize your figures.
- Time to discuss your purpose and intentions with professionals in your field and in illustration.
- Time to do your figures by computer or by hand or with an illustrator.

Learning of any kind requires patience:

- Mistakes and errors are inevitable.
- Experience cannot be rushed, and in the meantime the process itself can be enjoyable for you.
- Often the process of visualization and the making of visual choices can open your eyes and mind to further possibilities and challenges.

Practice is beneficial in many ways:

- The practice of focusing visually will sharpen your powers of analysis and observation.
- The practice of dealing with those who make figures for you will increase your knowledge of standards and methods of illustration.
- The practice of making your own figures will increase your efficiency and may open your eyes to other ways to illustrate.
- Practice can lead to experimentation, exploration, and growth.

Pictures are a natural, easy, and interesting way of learning and should be used as one of the essential ways to increase understanding. The writer Turgenev acknowledged in *Fathers and Sons* that "a picture shows me at a glance what it takes dozens of pages of a book to expound."

Bibliography

Abeloff, D., Medical art. Baltimore: Williams & Wilkins, 1982.

Alberts, B., Bray, D., Lewis, J., Raff, M., Roberts, K., Watson, J.D. Molecular biology of the cell. New York-London: Garland, 1989.

Alder, H.L., Roessler, E.B. Introduction to probability and statistics. San Francisco: Freeman, 1972.

Batschelet, E. Introduction to mathematics for life scientists, Volume 1. Berlin-Heidelberg-New York: Springer-Verlag, 1971.

Baylor College of Medicine. Handbook for effective visuals. Houston, Texas: Medical Illustration and Audiovisual Education, 1972.

Bojko, S. Diagrams, charts, graphs. Graphis 238. Zurich: Graphis Press, 1985.

Cardamone, T. Chart and graph preparation skills. New York-Cincinnati: Van Nostrand Reinhold, 1981.

Clayton, M. Leonardo da Vinci: The anatomy of man. Toronto: Bullfinch Press-Little, Brown & Co., 1992.

Cleveland, W.S. The elements of graphing data. Summit, N.J.: Hobart Press, 1994.

Council of Biology Editors. Illustrating science: Standards for publication. Bethesda, Md.: Council of Biology Editors, 1988.

Croy, P. Graphic design and reproduction techniques. London-New York: Focal, 1975.

Eastman Kodak. Communicating through poster sessions. Publication No. P-319, 1978.

———. Effective lecture slides. Pamphlet No. S-22, 1981.

———. Legibility—Artwork to screen. Pamphlet No. S-24, 1981.

———. Planning and producing slide programs. Publication No. S-30, 1981.

Edwards, B. Drawing on the artist within. New York: Simon & Schuster, 1987.

Herdeg, W. The artist in the service of science. Zurich: Graphis Press, 1974.

Herrlinger, R. History of medical illustration. New York: Medicina Rara, 1970 [English language translation].

Institute of Medical and Biological Illustration, ed. Charts and graphs. London: MTP, 1980.

Newman, J.R. The world of mathematics, Volume 1. New York: Simon & Schuster, 1976.

Philadelphia Museum of Art. Ars medica: Art, medicine, and the human condition. Connecticut: Meriden-Stinehour, 1985.

Saunders, J.B. de C.M., O'Malley, C.D. The anatomical drawings of Andreas Vesalius. New York: Bonanza Books, 1982.

Stryer, L. Biochemistry. New York: W.H. Freeman, 1995.

Trelease, S.F. How to write scientific and technical papers. Baltimore: Williams & Wilkins, 1958.

Tufte, R. The visual display of quantitative information. New York: Graphis Press, 1983.

Watson, J.D. Molecular biology of the gene. Menlo Park, Calif.: W.A. Benjamin, 1977.

Watson, J.D., Hopkins, N.H., Roberts, J.W., Steitz, J.A., Weiner, A.M. Molecular biology of the gene. Menlo Park, Calif.: Benjamin/Cummings, 1987.

Wood, P. Science illustration. New York: Van Nostrand Reinhold, 1983.

Wurman, R.S. Information anxiety. New York: Doubleday, 1989.

Zeiger, M. Essentials of writing biomedical research papers. New York: McGraw-Hill, 1991.

Sources of Illustrations

Page	Source
Chapter 1	
2–4	Wayne B. Lanier, Ph.D.
Chapter 2	
10 top	Tam K, Calonico LD, Nadel JA, McDonald DM. Globule leucocytes and mast cells in the rat trachea: their number, distribution and response to compound 48/80 and dexamethasone. Anat Embryol 178(2): 95–190, 1988.
10 bottom	Wright JR, Clements JA. State of art: metabolism and turnover of lung surfactant. Am Rev Resp Dis 135(1): 426–444, 1987.
11 top	Mary Helen Briscoe
11 bottom	Waldo Newberg, M.D.
12 top	Conhaim RL, Gropper MA, Staub N. Effect of lung inflation on alveolar-airway barrier protein permeability in dog lung. J Appl Physiol 55(4): 1249–1256, 1983.

PAGE	SOURCE
12 bottom	Wikman-Coffelt J. Biochemical regulation of myocardial hypertrophy. In A Zanchetti and R.C. Tarazi, eds. Handbook of Hypertension, Vol 7: Pathophysiology of Hypertension—Cardiovascular Aspects. New York: Elsevier Science BV, 1986.
13 top	Davis B, Nadel JA. New methods to investigate control of mucus secretion and ion transport in airways. Environ Health Perspective 35: 121-130, 1980.
13 bottom	Jay A. Nadel, M.D.
14	Waldo Newberg, M.D.

Chapter 3

19	Mary Helen Briscoe. Tracings courtesy of William H. Tooley, M.D., and Mureen Schleuter.
20	Mary Helen Briscoe
21	Wolfe CH, Moseley ME, Wikstrom MG, et al. Assessment of myocardial salvage after ischemia and reperfusion using magnetic resonance imaging and spectroscopy. Circulation 80(4): 969-982, 1989. By permission of the American Heart Association, Inc.
23	Michael A. Heymann, M.D.
24-27	Mary Helen Briscoe
28	Michael A. Heymann, M.D.
29	Mary Helen Briscoe. Photomicrographs courtesy of Robert L. Hamilton, Jr., Ph.D., and John P. Kane, M.D., Ph.D.
30	Mary Helen Briscoe
31-32	G. Chi Chen, Ph.D., and David Hardman
33-35	Mary Helen Briscoe. Echocardiogram courtesy of Nelson B. Schiller, M.D.

Chapter 4

38	Wayne B. Lanier, Ph.D.
39 top	Mary Helen Briscoe. Adapted from the following figure.
39 bottom	Henry Edmunds, M.D.

PAGE	SOURCE
40	Staub NC. Modern Physiology of Respiration. Chicago: Year Book Medical Publishers, Inc., 1989.
42	Bland RD, McMillan DD, Bressack MA. Decreased pulmonary transvascular fluid filtration in awake newborn lambs after intravenous furosemide. J Clin Invest 62(3): 601-609, 1978. By copyright permission of the American Society for Clinical Investigation.
43-45	Mary Helen Briscoe. Adapted from the preceding figure.
46	Mary Helen Briscoe. Created from data contained in the table on page 42.
47	Landolt CC, Matthay MA, Albertine KH, Roos PH, Wiener-Kronish JP, Staub NC. Overperfusion, hypoxia and increased pressure cause only hydrostatic pulmonary edema in anesthetized sheep. Circ Res 52: 335-341, 1983. By permission of the American Heart Association, Inc.

Chapter 5

PAGE	SOURCE
51	Jori Mandelman
52	Clive Pullinger
53-54	Mary Helen Briscoe. Adapted from Jori Mandelman.
55 top	Jori Mandelman
55 bottom	Mary Helen Briscoe. Adapted from the preceding figure.
56	Clive Pullinger
58 top	Mary Helen Briscoe. Adapted from Jori Mandelman.
58 bottom	Jori Mandelman
59	Mary Helen Briscoe. Adapted from the preceding figure.
60	Jori Mandelman
61-63	Clive Pullinger
64	Jori Mandelman
65 top	Chuen-Shang C. Wu, Ph.D., and Jen Tsi Yang, Ph.D.
65 bottom	Clive Pullinger

PAGE	SOURCE
66	Adapted from Richardson JS. Anatomy and taxonomy of protein structure. Adv Protein Chem 34: 167-339, 1981. For Chuen-Shang C. Wu and Jen Tsi Yang.
67-69	Mary Helen Briscoe

Chapter 6

PAGE	SOURCE
74	Mary Helen Briscoe
75	Mary Helen Briscoe. Adapted from data from Terjung RJ, Baldwin KM, Winder WW, Holloszy JO. Glycogen repletion in different types of muscle and in liver after exhausting exercise. Am J Physiol 226(6): 1387-1391, 1974.
76	Mary Helen Briscoe. From Terjung et al., op. cit., p. 67
77	Mary Helen Briscoe
78	Mary Helen Briscoe. From Terjung et al., op. cit., p. 67.
79	Mary Helen Briscoe
80-82	Mary Helen Briscoe. From Terjung et al., op. cit., p. 67.
83-84	Mary Helen Briscoe
85	Mary Helen Briscoe. From Terjung et al., op. cit., p. 67.
86 top left	Mary Helen Briscoe. From Terjung et al., op. cit., p. 67.
86 top right and bottom	Mary Helen Briscoe
87	Mary Helen Briscoe. From Terjung et al., op. cit., p. 67.
88, 89 top	Mary Helen Briscoe
89 bottom, 90-96	Mary Helen Briscoe. From Terjung et al., op. cit., p. 67.
97-100	Mary Helen Briscoe
101	Mary Helen Briscoe. From Terjung et al., op. cit., p. 67.

Chapter 7

PAGE	SOURCE
105-111	Mary Helen Briscoe
112-113	Mary Helen Briscoe. From Terjung et al., op. cit., p. 67.

PAGE	SOURCE
Chapter 8	
117–119	Mary Helen Briscoe. Modified from A guide to illustrations for medical and scientific research. Regents, University of California, 1986.
120–121	Mary Helen Briscoe
122–124	Mary Helen Briscoe. From Terjung et al., op. cit., p. 67.
126	Mary Helen Briscoe
127	Hotkins AL, Koprowski H. How the immune response to a virus can cause disease. Sci Am 228(1): 22–31, 1973.
Chapter 9	
132–133	Mary Helen Briscoe
135 top	From meeting instructions to authors.
135 bottom, 136–140	Mary Helen Briscoe
141	Baker DG, Don H. Catecholamines abolish vagal but not acetylcholine tone in the intact cat trachea. J Appl Physiol 63(6): 2490–2498, 1987.
142–143	Mary Helen Briscoe. From Terjung et al., op. cit., p. 67.
144	Mary Helen Briscoe
Chapter 10	
152, 154	Mary Helen Briscoe
Chapter 11	
159	Mary Helen Briscoe
160	Mary Helen Briscoe. From Wayne B. Lanier, Ph.D.
161, 162 top	Mary Helen Briscoe
162 bottom, 163 bottom	Morales M. The search for a biological energy transducer. Science as a way of knowing. Am Zool 29: 593–603, 1989.
164	Joan Wikman-Coffelt, M.S., Ph.D.
166–167	Mary Helen Briscoe

PAGE	SOURCE
Chapter 12	
171–176	Mary Helen Briscoe
177	Chuen-Shang C. Wu, Ph.D., and Jen Tsi Yang, Ph.D.
178, 182	Mary Helen Briscoe
Chapter 13	
184 top	Courtesy of the Proctor & Gamble Company.
184 bottom	Guttentag OE. The attending physician as a central figure. In EJ Cassell and M Siegler, eds. Changing Values in Medicine. Frederick, Md.: University Press of America, 1979.

INDEX

S